SIMPLE BOAT MAINTENANCE

Pat Manley

FERNHURST
BOOKS

www.fernhurstbooks.com

This edition first published in 2014 by Fernhurst Books Limited
62 Brandon Parade, Holly Walk, Leamington Spa, Warwickshire CV32 4JE
Tel: +44 (0) 1926 337488
www.fernhurstbooks.com

First published in 2005 in hardback by Fernhurst Books
Reprinted in 2011 in hardback by John Wiley & Sons Ltd

British Library Cataloguing in Publication Data
A catalogue record for this book is available from the British Library
ISBN 978-1-909911-13-0

Acknowledgments
The author would like to thank the following people and organisations for their generosity and help:
Colin Bridle for allowing him to photograph the section on repairing a punctured hull.
Peter Spreadborough of Southampton Calor Centre for the use of his photographs of gas joints and
fittings.
Jeff Sheddick of ITT Jabsco for supplying Jabsco products for photography.
Shamrock Chandlery for allowing him to photograph some of their stock items.
Blakes, Volvo Penta, Perkins Sabre Engines, Vetus, Lofrans and Lewmar for information, diagrams and
photographs.

Photographic credits
All photographs and diagrams are by the author except:
Peter Spreadborough: pages 65-69; Blakes Lavac Taylors: page 111; Volvo Penta, diagrams and photos:
pages 16, 20, 27, 38, 46, 48; Vetus: page 117; Perkins, reprinted courtesy of Perkins Engine Co. Ltd: page
25; Lewmar: pages 120, 121, 157; Lofrans: page 157; Harken: pages 1, 151

Cover design by Rachel Atkins
Design and artwork by Creative Byte & Rachel Atkins
Printed in China by Toppan Leefung

CONTENTS

FOREWORD

This book has been a long time in gestation, with many thoughts on the best way to present the information. Tim Davison, my publisher, has guided me and 'Simple Boat Maintenance' is the result.

Without my wife, Lynette, this book would not have been possible. Either her hands are shown in the photographs doing the jobs or she has taken the photographs. She has fed and watered me while I was sitting at my computer and soothed my brow when the computer was being beastly.

With this book in one hand and a spanner in the other, I hope you will enjoy trouble-free boating.

Pat Manley, Hythe, Southampton

INTRODUCTION

Lots of jobs on your boat are relatively straightforward and don't need a great deal of skill. However, if you haven't tackled the job before, you may not know the techniques required, or know how to set about the job.

Simple Boat Maintenance aims to talk you through much of the maintenance on your boat so that you don't have to incur unnecessary and expensive labour charges. I have done all these jobs myself, so any of them should be within your capability. For each job the skill level required is shown on a scale of one to five to give an indication of difficulty (five is probably a job for a professional). No special tools are required.

Because there are many different products of each type on the market, I can't show the work involved on every one of them. However, armed with *Simple Boat Maintenance* and the instructions that come with the product, you should be able to complete any of the work shown in this book successfully.

Jobs that require an above average level of skill are not covered in this book. But you should be able to reduce the cost of ownership and increase the level of enjoyment of owning your boat by using this book as a guide.

Each topic is covered under the following headings:

- WHY IT NEEDS TO BE DONE
- WHEN IT SHOULD BE DONE
- SKILL LEVEL REQUIRED
- TOOLS NEEDED
- WHERE TO FIND IT
- HOW TO DO IT

Sometimes a topic has information only, and in this case the page is coloured yellow.

Routine maintenance

This is carried out regularly, often on an annual basis, and doesn't come as a surprise. You can choose when to do it, and have any tools and spares to hand ready for the job. If the boat gets a lot of use, the time interval between jobs may need to be shortened. The equipment's instructions or handbook will tell you what's needed and when. But do note down what you do and when you do it, maybe in the boat's logbook. This will keep you informed and should be a good selling point if you decide to sell your boat.

Make a list of all the routine maintenance required on your boat and tick off each job as it's done.

Mending something that has developed a fault

Sometimes it is patently obvious that something has broken and all you need to do is replace it.

At other times troubleshooting may be required to find out what's wrong and how to cure the problem.

A philosophy of troubleshooting

When things are all right, they are *NORMAL*. So the first thing we need to do is to establish what this normality is. Seems simple really, doesn't it, yet as things change slowly we may lose a sense of what is normal for that particular item.
For instance: the engine may take longer to start; the genoa winch may be stiffer to operate; the helm may be heavier than it used to be.

Sometimes normality is easy to establish, such as the engine's normal running temperature. Maybe it is normally 70°C, but we don't have to remember this: a little paint-mark on the instrument where the needle usually resides will do the job

nicely. If we have two engines, why have the temperature gauges on opposite sides of the panel? Put them next to each other and any change is obvious. Unfortunately, designers are often more interested in style rather than doing a proper job!

We need to have a feeling for what is *NORMAL*. This allows us to know when something has changed and it is *CHANGE* that tells us that something may be wrong.

Then we need to have a look at the change to see if it's real or significant. For instance, the fact that the boat is going more slowly than usual at a given rpm doesn't have to indicate that the engine has lost power. It may mean that the hull or propeller is fouled or even that the log impeller is fouled. So we have to look at *ALL* the symptoms. In this case a glance at the wake of the boat would tell us if it were a fouled impeller. The fact that the hull was last antifouled a year ago would indicate the strong likelihood that fouling was the cause. Only having eliminated these would we turn our attention to the engine.

Basically, then, if we believe we have a problem, we need an orderly approach to curing it.

- What is normal?
- What has changed?
- Look at all the symptoms.
- What likely cause(s) matches as many of the symptoms as possible?
- List these possible causes in the order of least cost or, most likely, ease of checking.
- Take some specialist advice if necessary.
- Decide on a plan of action.
- Tackle the job.

But, whatever maintenance you have to do, enjoy your boating!

ENGINES: Cooling system

The majority of the energy contained in the fuel is wasted. At least 25% of this waste goes into the engine cooling system and must be removed, otherwise the engine will overheat and serious engine damage will occur. Any failure within the cooling system may cause a very rapid rise in engine temperature, requiring the engine to be shut down.

Often the first indication of overheating is the overheat alarm. If a temperature gauge is fitted this will give earlier warning, but it's often overlooked. If overheating is caused by failure of the seawater pump, or blockage of its supply, cooling water will fail to flow from the exhaust, causing a change in the exhaust note. The earlier this is detected, the greater the chance of avoiding failure of the seawater cooling pump impeller.

There are two forms of water cooling:
- *Direct (raw water) cooling*
  The engine is cooled by the water in which the boat floats. So, if you are on the sea, seawater circulates through the engine's cooling system.
- *Indirect cooling*
  Fresh water circulates around the engine's cooling system and is cooled by a heat exchanger. Usually seawater (or river water) flows through the heat exchanger to remove the heat. On some boats a keel cooler, mounted outside the hull and immersed in the sea or river water, carries the heat away (see diagrams below).

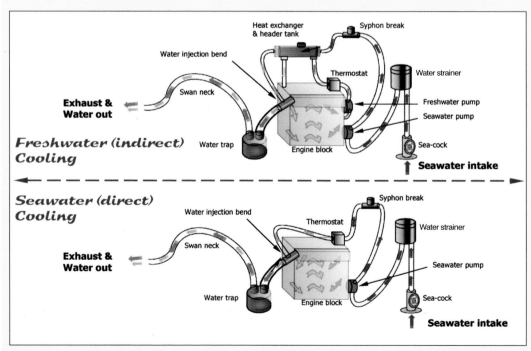

Freshwater (indirect) Cooling

Heat exchanger & header tank • Syphon break • Water injection bend • Thermostat • Water strainer • Swan neck • Freshwater pump • Seawater pump • **Exhaust & Water out** • Water trap • Engine block • Sea-cock • **Seawater intake**

Seawater (direct) Cooling

Syphon break • Water injection bend • Thermostat • Water strainer • Swan neck • Seawater pump • **Exhaust & Water out** • Water trap • Engine block • Sea-cock • **Seawater intake**

Direct cooling

This is the simplest system – but with seawater flowing through the engine the waterways will suffer from corrosion. Many older engines are made from cast iron and have thick walls and so have a long life expectancy. Some modern light-weight engines also have direct cooling and may have sacrificial anodes to help control corrosion. It's essential that you check the engine handbook to find out where they are and change them annually at least.

The thermostat controls how much seawater flows through the engine to control its running temperature. When the engine is cold, no water passes through the engine but flows directly into the exhaust. As the engine heats up some cooling water is directed through the engine's waterways and then recombines with the main flow before being injected into the exhaust system. If the engine starts to run too hot, all the cooling water flows through the cooling ways.

1 Engine overheating due to blockage

WHY IT NEEDS TO BE DONE
Engine overheating can cause severe damage to the engine.

WHEN IT SHOULD BE DONE
If there is an indication of overheating. Overheating may be indicated by an over-temperature alarm or a higher than normal reading on the temperature gauge, if fitted.

TOOLS NEEDED
Spanners and screwdrivers.

WHERE TO FIND IT
Anywhere, from cooling water intake to exhaust.

HOW TO DO IT
Unless the cause is obvious, you need to work your way around the system, starting at the most likely or the easiest to achieve.

Difficulty Rating: ▨■□□□

(Knowledge of how a cooling system works is invaluable for troubleshooting.)

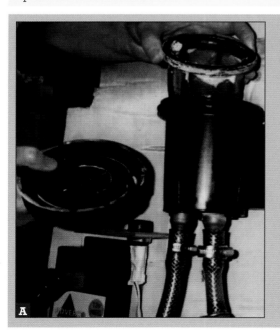

A

How do I know it's a blockage?

1. If you suspect a blockage, stop the engine.
2. Wait a minute and restart. (With luck this will allow any plastic sheeting or bag, which has been sucked up against the seawater intake, to float clear).
3. Observe the exhaust or listen to it to check if there's a flow of cooling water.
4. If not, stop the engine again and investigate further.
5. Check the seawater strainer to see if the filter is blocked. Clean it if necessary (photo A).
   Some builders don't fit a seawater strainer when installing a Volvo saildrive. If yours hasn't got one, fit one in an accessible place.

Indirect cooling

All the seawater flows through the heat exchanger all the time. None flows through the engine itself

A circulation pump circulates the fresh water round the engine's waterways. The thermostat controls how much of the fresh water flows through the fresh water part of the heat exchanger to maintain the engine at the desired temperature.

No anodes are required in the fresh water part of the system, BUT there are often anodes in the heat exchanger itself. Where several heat exchangers are used, for cooling the oil, gearbox or turbocharged air, there MAY be an anode in each heat exchanger.

Seawater strainer

It's desirable to fit a strainer in the seawater system prior to the seawater pump. This will prevent weed and other marine life from entering the engine's cooling system. It will also help prevent silt entering the system. Many older boats didn't have one fitted and, even now, some boat builders don't bother to fit one either. If you haven't got one, fit one.

B

6 A blockage between the seacock and the strainer can often be cleared by means of a dinghy inflation pump or a gas foghorn (photo B). You'll probably have to disconnect the hose from the inlet to the strainer so you can blow down the pipe.

C

D

7 If the strainer is below the waterline, you'll need to close the seacock first and then reopen it as you blow (photo C).

8 On saildrive engines, the seacock will be found close to the gearbox, (port side on Volvo Pentas and starboard on Yanmars). Early Volvos have a plastic wheel (photo D), which turns through only 90 degrees and is very stiff. Later engines have a lever-operated seacock instead. Yanmars have a brass screw-type cock with a 'T' bar handle that needs several turns to close it off. On shaft-drive engines you will have to follow the inlet hose until you find the seacock.

2 Seawater pump

Where's the seawater pump?

1 On many smaller engines the seawater pump is located on the front of the engine, where generally it's accessible (photo E1). However, that's not always the case.

2 Older Volvo Pentas have them above the gearbox (photo E2).

3 Follow the pipe from the seawater strainer and you'll come to the pump.

TIP — Yanmar 2GM and 3GM engines

On the Yanmar 2GM and 3GM the seawater pump, although on the front, faces backwards alongside the front starboard engine mounting. The easiest way to get at the impeller is to remove

the pump first (photos F1 & F2). This isn't as bad as it sounds because only two bolts need to be removed and there's no need to disconnect the pipes.

Causes of pump failure

- If the pump runs dry it will overheat and the blades will disintegrate. This is the most common cause of failure (photo A).

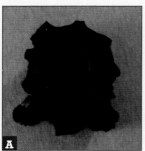

- Pieces of blade can lodge in the waterways causing a partial blockage and will then cause further overheating. In seawater-cooled engines (direct cooling) the debris could end up almost anywhere and may or may not cause a further problem. Back-flushing the cooling system may cause the debris to return to the pump body, so if you are concerned you may well have to consult an engineer.

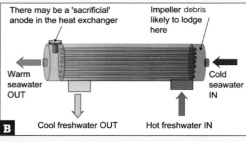

There may be a 'sacrificial' anode in the heat exchanger

Impeller debris likely to lodge here

Warm seawater OUT   Cold seawater IN

B   Cool freshwater OUT   Hot freshwater IN

- With fresh water-cooled engines (indirect cooling) debris will go directly to the heat exchanger and either lodge there or pass right through into the exhaust (diagram B). Cleaning the heat exchanger tubes is a comparatively simple task.

- Pump impellers sometimes become unbonded from their central metal boss (photo C). The impeller then remains stationary in the

pump while the boss drives round. When you remove the faceplate this is not at all obvious, so when you extract the impeller try turning the impeller while you firmly grip the boss to check the bonding. Several batches of impeller suffered from this problem so this can occur very quickly, even with a new one.

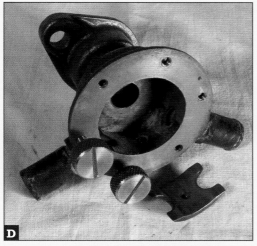

- Wear of the pump cam (photo D) will cause a reduced rate of flow. This reduced flow may be insufficient to cool the engine and overheating can then occur.
- An even rarer problem is for the screw holding the wedge-shaped cam inside the pump case to shear. This cam then rotates with the impeller rather than staying stationary so check that the cam

is between the inlet and the outlet the 'short way round' (photos E and F). If the screw has sheared you'll probably have to remove the pump to replace the screw. The replacement must not be so long as to stand proud of the cam or the impeller will be damaged. The pump often needs to be removed to access the screw (photo G).

- If there is significant wear on the front plate (photo H), as a temporary measure it can be reversed so that the front side faces the impeller (photo I). Any inscription will not affect the pump's performance.

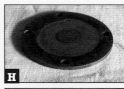

Changing the impeller

WHY IT NEEDS TO BE DONE
Failure of the impeller will cause overheating of the engine.

WHEN IT SHOULD BE DONE
The impeller needs to be changed annually as a precaution against failure. Blockage of the cooling water intake will cause the impeller to overheat and fail.

TOOLS NEEDED
Spanners and screwdrivers.

WHERE TO FIND IT
Often on the front of the engine, but it can be on the side or rear.

Difficulty Rating:

Changing the impeller

1. Turn off the seawater cock (photo J).
2. Remove the pump faceplate securing screws (diag. K).

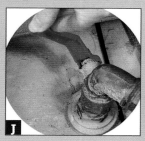

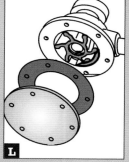

3. Remove the faceplate (diag. L).
4. Remove the impeller using longnosed pliers, slip pliers or an impeller removal tool (Jabsco) (diag. M).

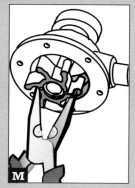

5. Inspect the old impeller for wear and cracks (photo N).

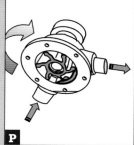

6. Insert the new impeller (photo O).
7. Impeller revolves 'long way round' from inlet to outlet (diag. P).
8. Replace faceplate and gasket (renew the gasket if necessary).
9. Insert all screws and tighten by hand.

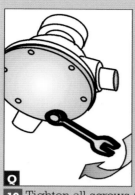

When you restart the engine, check the exhaust to make sure that cooling water is coming out, then check the pump faceplate joint for leakage. If the faceplate is hot to the touch, it's running dry. Make sure that you don't get caught up in any moving parts of the engine or shaft.

One yachtsman had a cooling water failure and on removing the faceplate and impeller a small eel dropped out into the bilge. You never know what you might find!

10 Tighten all screws (diag. Q).
11 Re-open the seawater cock (photo R).

TIP

Inserting the new impeller

The new impeller may be inserted either face in, but the blades should be deflected in the correct rotational direction as you insert it. If you bend the blades into place by tightening a cable tie, piece of string or a strong rubber band round the impeller, you'll find the job of inserting it much easier (photo S).

TIP

Fit a SPEED SEAL

The 'SPEED SEAL' by True Marine (photo T) replaces the standard faceplate and gasket with a heavy duty plate and an 'O' ring seal. These are attached by large knurled screws, two of which remain captive in the pump body, which need only be finger tight. The advantages are that you are less likely to drop small parts into the bilge and you don't need any tools to remove the faceplate. Speed Seals are available for most small engines.

Seawater pumps on Sabb engines

Sabb engines have a very different type of pump, employing a diaphragm (photos B1 and B2). This is much more complicated but can be run 'dry' for considerable periods without failure. They compare well with impeller pumps, which will run for a maximum of 10 minutes without water to lubricate them.

Seawater pump leaking

WHY IT NEEDS TO BE DONE

The water and oil seals on the water pump shaft will wear with time, leading to pump shaft corrosion or loss of engine oil.

WHEN IT SHOULD BE DONE

A daily check of the pump's 'tell-tale' hole will give advance warning of seal failure.

TOOLS NEEDED

None.

WHERE TO FIND IT

The water pump is often mounted on the front of the engine. However on some it's mounted on the rear or the side, so check your handbook.

Difficulty Rating: ☐☐☐☐☐

Leaks from the pump body

1 There's a water seal on the drive shaft and possibly an oil seal as well. The water seal, especially, will start to leak after a while and if this isn't replaced fairly quickly, shaft damage will occur. If there's an oil seal it will usually be identical to the water seal, but fitted facing in the opposite direction.

2 The pump body will have a 'tell-tale' drip hole on its underside (photos A and B) and any leakage past the seals will be detected here. A daily wipe of the underside of the pump body with a finger will quickly detect the first signs of a leak. You can continue to use the engine but the sooner you do something about it the cheaper the repair will be.

3 Replacement of the seals is beyond the scope of this book but many

Pump 'tell-tale' drip hole

A

B

pumps have a repair kit available. You may need to enlist the help of your dealer to remove and replace the shaft bearings to allow the seals to be changed.

TIP Some engines have no repair kit available so eventually a replacement pump will be required. It's often possible to get a different make of pump to replace your old one at half the cost of the original – and it may have spares available as well. Check with equipment suppliers.

Leaking from the faceplate

1 You won't normally get a leak from the faceplate unless you've just put it back, although in time weeping round the edges may become apparent. (photo C).

C

2 Remove the faceplate and check the gasket. These are normally very thin and easily damaged.

D

3 A smear of waterproof grease in conjunction with the gasket will probably keep the leak at bay for a while (photo D).

4 If the gasket is too badly damaged and you haven't got a spare, an application of a silicone gasket sealant (such as Locktight 5926 flange sealant) will do the trick (photo E).

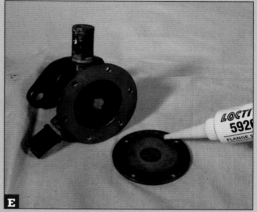

E

5 If you use a sealant, fully tighten the screws immediately, without waiting for the sealant to set (as you normally would). This ensures minimum gasket thickness and thus maximum impeller compression.

F

G

6 If you have fitted a 'Speed Seal' (photo F), check the rubber 'O' ring seal and replace if necessary (photo G).

3 Thermostat

Thermostat failure

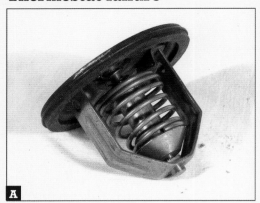

A

B

- Modern, wax filled thermostats (photo A) are generally reliable. If they do fail, they tend to do so in such a way that they will cause the engine to run too cool, rather than too hot. This has efficiency implications but will not cause the engine to overheat.
- Older types, which are still found in many engines, have an air-filled bellows (photo B) and when these fail they will normally cause the engine to

overheat. Many sources tell you that you can remove the thermostat as a get-you-home measure. This is not generally true and don't attempt to do it. If you've no spare, you'll need to immobilise it in a partially open position to get you home. The engine will probably run too cool, but could overheat in extreme conditions, so watch the temperature gauge if you have one.

TIP

Testing a thermostat

F

The thermostat can easily be tested for correct operation. If you put it in a saucepan of water and heat it on the stove, you can see it open as the temperature rises. A thermometer will indicate correct operation as the opening temperature will be stamped on the thermostat. In many cases the thermostat is changed unnecessarily in the hope that it will cure a problem, when a simple check would have revealed that it was, in fact, serviceable (diagram F).

Removing the Thermostat

WHY IT NEEDS TO BE DONE
If you believe the thermostat is the cause of overheating check it or replace it.

WHEN IT SHOULD BE DONE
As required in the troubleshooting process.

TOOLS NEEDED
Spanners and screwdrivers.

WHERE TO FIND IT
Look for the thermostat housing (often at the end of a flexible pipe) just after the seawater pump. It's normally held in place by a couple of bolts.

Difficulty Rating:

Removing the Thermostat

1 Turn off the cooling water seacock.
2 Loosen the seawater pump faceplate to drain water from the thermostat housing. (Not essential, but stops seawater flooding onto the engine.)
3 Loosen the pipe clamps on any inlet and outlet pipes on the thermostat housing.

B

A

4 Remove the bolts attaching the thermostat housing (photo A).

5 Lift the housing clear (photo B).

C

6 Lift out the thermostat (photo C).
7 Test or renew the thermostat and replace it in the housing.
8 Reassemble in reverse order, renewing the gasket if neccessary.
9 Open the cooling water seacock.
10 Restart the engine, check the cooling water flow and check for leaks. The housing will remain cool until the thermostat opens. Then it will become hot to touch as hot water starts to flow inside it.
11 Recheck the cooling system symptons.

4 Anti-syphon valve

On sailing boats the exhaust pipe is swept up into a 'swan neck' to stop following waves flooding the exhaust when the engine isn't running. Exhaust gas normally drives water up over the swan neck and out of the system, but of course if the engine isn't running and seawater is entering the system, water will accumulate in the lowest section of the exhaust pipe. Eventually there will be enough to enter the exhaust manifold and run into any cylinder that has its exhaust valve open.

- Water can enter the exhaust when water syphons past the water pump because the exhaust injection bend is below the waterline (diagram B).
- Water entering the working parts will contaminate the oil and cause serious corrosion (photo A).
- Water is incompressible, and can shatter the piston or bend a connecting rod if you attempt to start the engine with water in the cylinder.
- To prevent water from syphoning into the engine fit a syphon break if the water injection point into the exhaust is below the waterline. Because of heeling and pitching, syphoning could occur with the water injection point as much as 150mm above the static waterline.

Important note:
On a sailing boat and some motor boats water can be trapped in the exhaust system if you continue to turn the engine on the starter motor and the engine doesn't fire.

The seawater pump drives water into the exhaust but there's no exhaust gas to drive it out.

If you must continue to crank the engine, first close the cooling water intake seacock.

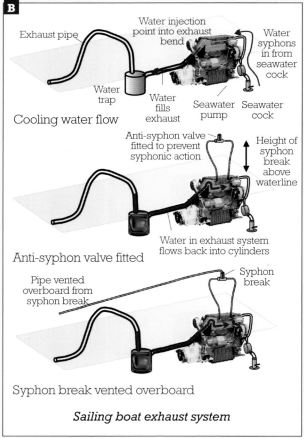

Cooling water flow

Anti-syphon valve fitted

Syphon break vented overboard

Sailing boat exhaust system

WHY IT NEEDS TO BE DONE

An anti-syphon valve will suffer from a build-up of salt crystals, blocking the air valve and causing it to become inoperative. An anti-syphon valve is fail-dangerous, in that you know neither if it's working or if it has failed.

WHEN IT SHOULD BE DONE

Annually.

TOOLS NEEDED

Screwdriver, spanner.

WHERE TO FIND IT

- If one is needed, the builder will probably have installed one. You may not know it's there as it needs to be fitted at least 150mm above the waterline so it's often out of sight.
- To find out if you have one, look to see if the seawater supply pipe from the pump runs upwards above the engine and then returns down before going into the cooling system. You might instead find it doing the same thing, but after it leaves the engine prior to entering the exhaust (photo C).

The Vetus anti-syphon valve shown in photo C comes in two varieties. The syphon break, as in the photo, and the anti-syphon valve, which has no tube.

Be aware that sometimes a short length of tube is led to the bilge from an anti-syphon valve to vent any dribbles of water that may occur. This is still an anti-syphon valve and needs to be serviced annually.

- The ideal system uses no valve, but has a small diameter tube draining overboard well above any possible heeled waterline. This is a 'fail-safe' system and requires no maintenance except to observe that there's a flow of water out of it when the engine's running (photo D).

C

- You can tell when there's a valve because there's no tube running overboard from it, although there may be a short length dropping down towards the bilge.

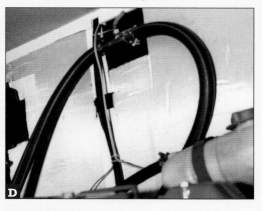
D

Difficulty Rating:

Cleaning an anti-syphon valve

1 Remove the top from the valve assembly.

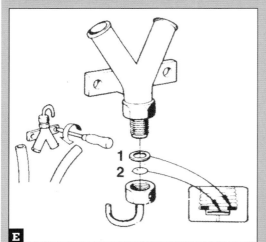

E

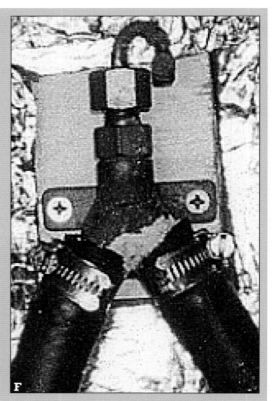

F

2 Carefully remove the valve and clean it and the seating (diagram E).

3 Check that the vent pipe is clear by blowing through it.

4 Carefully reassemble the unit.

 TIP   **Volvo Penta anti-syphon valve**

This unit must be held 'upside down' when you reassemble it. It's often much easier to remove this valve from the bulkhead when you service it, unless it's very accessible (photo F).

5 Other problems

Overflowing coolant

1 If the heating system is overfilled, water will expand and overflow as the engine heats up. Once the correct level is attained, overflowing will cease.

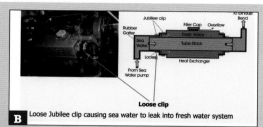

2 Engines with rubber seals at the ends of the heat exchanger can suffer from continual over-flowing (photo A).

3 This is usually caused by a loosened jubilee clip allowing water to leak from

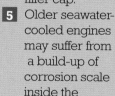

Loose Jubilee clip causing sea water to leak into fresh water system

the seawater system into the fresh water system. This will allow the fresh water to become contaminated and corrosion will occur (photo B).

4 Tighten the jubilee clip and flush and refill the fresh water system.

Further troubleshooting

1 If problems of overheating still exist, much more thought will have to be given to the problem, as there are no clear-cut answers.

bleed screw

2 In a fresh water-cooled system (indirect cooling) an air lock could be the cause and, where this is likely, the handbook should detail the bleeding procedure. This is particularly true of the Perkins 4107 and 4108 engines, which have bleed screws in the cylinder head (photo C).

3 After refilling the cooling system, you will probably need to run the engine with the cooling water cap removed until the engine is hot, topping up the system as the level falls on the expulsion of any air (photo D).

4 Once the water level is stable, fit the cap, fill the expansion chamber, (if fitted) to the hot level and stop the engine. When the engine is cool check the level in the exp-ansion chamber (photo E) or the filler cap.

5 Older seawater-cooled engines may suffer from a build-up of corrosion scale inside the waterways, causing a restriction of flow.

6 A proprietary de-scaler, such as Volvo Part number 1141658-3 may be used, but follow the instructions explicitly to avoid engine damage.

7 A build-up of marine growth in the sea-water inlet can also cause a restriction. Once the obvious causes have been investigated without success, it's probable that professional help will have to be enlisted as the causes go beyond the scope of simple maintenance.

Fuel system schematic

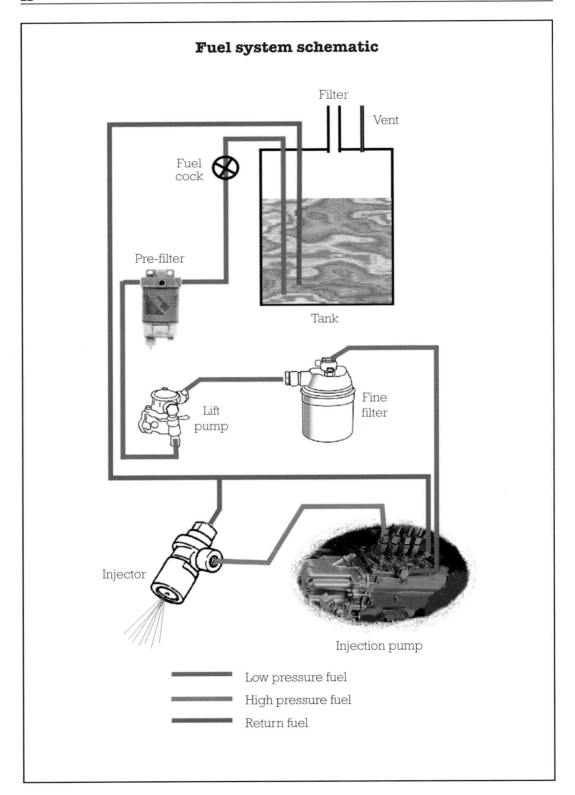

ENGINES: Fuel System

6 Fuel pump

The fuel pump transfers fuel from the tank to the engine at low pressure. On small engines it's usually driven directly by the camshaft (photo A). On larger engines, it may be part of the fuel injection pump (photo B). See also schematic opposite.

WHY IT NEEDS TO BE DONE

Fuel pumps are very reliable, but if they fail the engine won't run – unless the level of fuel in the fuel tank is higher than the engine's fuel injection pump.

WHEN IT SHOULD BE DONE

Older engines may have a pump with a built-in filter, in which case it should be cleaned annually. Otherwise the pump needs attention only if a problem develops.

TOOLS NEEDED

Usually a set of Allen keys (hexagonal wrenches).

WHERE TO FIND IT

Often low down on the side of the engine. It may not be easily accessible (photo A). However, some engines incorporate the fuel lift pump in the fuel injection pump (photo B). If these fail, it's a job for a company specialising in fuel injection.

Difficulty Rating:

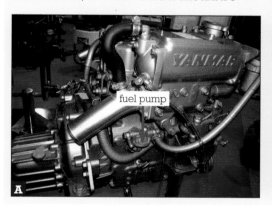

fuel pump

A

hand primer

fuel pump

B

On older engines the fuel lift pump can be taken apart and repaired. Newer ones cannot, they're a throwaway item. So what can go wrong? How, indeed, do you know it's failed?

Is the pump working?

- If the level of fuel is above the highest part of the fuel system, the engine is being gravity fed and the pump is doing nothing. A pump failure in this case will pass unnoticed.
- If fuel starvation is found to occur at a specific level of fuel contents then suspect pump failure.
- If the system requires fuel to be lifted at all times, pump failure will prevent the engine from running at any time. ➤

Checking the pump

1 Older pumps have a lid, which can be removed, revealing a filter (photos A to E).

A

B

C

D

E

2 If the lid is stuck, lever it off with the bolt (photos B and C).

3 Check the filter and clean if necessary (photo F).

F

4 Ensure that there's fuel in the tank and the fuel cock is open.

5 Open the bleed screw on the engine filter.

6 Operate the hand-priming lever or plunger on the fuel lift pump (photo G).

G

7 If no fuel emerges after a reasonable time, close the bleed screw.

8 Pump failure is likely, but see note below.

Note:

- Some builders have installed an unadvertised filter on the fuel pipe inside the tank. This is often inaccessible, and should it become blocked no fuel will flow from the tank.
- An air leak on the suction side of the pump will prevent fuel from being withdrawn from the tank – unless the fuel level is above the pump.

Older pumps

1 Older pumps can be taken apart for servicing (diag. A).

2 There are many different types and unless you have the service manual, it's best left alone.

3 You'll need to remove the pump to service it (photos B to F).

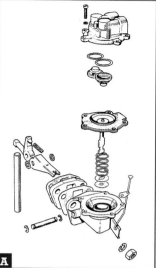

A

E

F

Newer pumps

- All you can do with the newer type of pump is replace it.

G

- Often the replacement will look different, especially if you are replacing an obsolete pump (photo G).

Getting home with a failed fuel pump

- Rig up a temporary gravity feed system (diagram H).

B

C

D

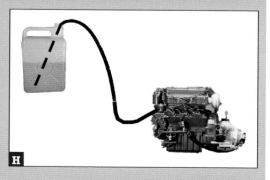

H

7 Bleeding (venting) air from the fuel system

If air gets into the fuel system the engine will run erratically with the engine speed alternately falling and rising and it will eventually stop. If you manage to start it, it will run at only low rpm and again will eventually stop.

Air can enter the fuel system because the tank is empty, or a low level of fuel causes the intake pipe to cover and uncover as the boat pitches and rolls. If there is a leak between the tank and the fuel pump, above the level of the fuel tank, air will enter the fuel system. Air will also enter if you have changed the fuel filters or undone any pipe joints.

Difficulty Rating: ▓ ▓ ▢ ▢ ▢

Bleeding the fuel system

Once the cause of air entering the system has been removed (see below), bleed the system to remove any air.

1 Open the fuel cock.

2 If the water separating pre-filter (or agglomerator) is below the level of the fuel in the tank, open its bleed screw about 2½ turns (photo A).

A

If in doubt of the fuel level, don't open this bleed screw and go to step 4.

3 Wait until fuel issues freely from the bleed screw and then close it. (You'll get air, then bubbles, then pure fuel – check that it is fuel, not water. Fuel feels slippery to the touch, water doesn't.)

4 Now open the bleed screw on the engine fuel filter (photo B).

5 Pump the priming lever on the fuel lift pump until fuel flows from the bleed screw.

B

6 Close the bleed screw and continue pumping a few more times. The engine should now start.

TIP **Finding a fuel leak in a pressure pipe**

Fuel spreads from the site of the leak when the engine is running, so pinpointing the leak may be difficult.

1. Dry everything well. (Bicarbonate of Soda works well to remove the fuel.)
2. Use a 'puffer' to apply talcum powder to a suspect area.
3. Start the engine.
4. Darkening of the powder should help locate the leak.

TIP **Pump won't prime**

If you find the hand priming lever feels dead with no 'spring' to it, give the starter a 'blip'. The engine had probably stopped with the pump's operating cam at 'top dead centre' so that the pump's stroke was at maximum. In this position, operating the hand-priming lever will have little or no effect.

Removing air from high pressure pipes

Very occasionally you may need to remove air from the high-pressure pipes joining the fuel injection pump to the injectors.

1 There are no bleeding points so you'll need to loosen the joints at the injectors (photo C).

C

2 This fuel is under high pressure (around 200 atmospheres) and can puncture the skin or eyeball so put some cloth around the joint as a protection.

3 The engine will have to be turned over on the starter motor to operate the fuel injection pump. The engine could start, so be sure to keep clothing and body away from any rotating parts.

4 If no fuel comes out, check that the fuel stop handle (if fitted) is fully home or the fuel stop solenoid is at run. (Diag. D & photo E.)

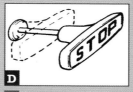

D

E

5 Fully tighten the pipe joints when you have finished.

 TIP Identifying the bleeding points

Paint the bleed screws with paint in a contrasting colour (photo F). Also keep any necessary 'emergency' tools securely fastened in the engine compartment where they can be found in a hurry. Ban their use for other jobs.

F

 TIP Finding a leak in a suction pipe

A leak in the pressure side of the system, or where the leak is below the level of fuel in the tank in the suction side, can be found because fuel will come out of the leak. Where there's a leak in the suction pipe but above the level of fuel in the tank, air will enter and it's difficult to find.

You need to try and pressurise the tank, but to do this you'll have to block off the tank filler and vent pipes while at the same time you introduce air using a dinghy inflation pump. How you achieve this will depend on the installation and its accessibility. A combination of softwood plugs, jubilee clips and a piece of pipe may well do the job.

Having pressurised the tank a little, fuel should come out of the leak. If the pipe is empty, then paint a solution of washing up liquid and water on any suspect joint or area and air leaking out will cause the solution to bubble.

ENGINES: Electricity

8 Drive belts

WHY IT NEEDS TO BE DONE

A rubber drive belt is used to drive the alternator (or on older engines the dynamo). Correct tension is important – too tight and bearings will suffer, too loose and slip will occur. Slip will cause overheating of the belt, leading to its premature failure and also, where the tachometer is powered from the 'W' connection on the alternator, rpm will under-read. A broken drive belt will stop battery charging, but the engine will run.

On some engines the water pump is also belt-driven. Failure of the drive belt will cause rapid overheating of the engine and prevent continued running.

Belt pulley alignment is also important for the longevity of the belt, although this will normally be a problem only if pulleys or the alternator have been changed.

WHEN IT SHOULD BE DONE

A quick check of belt tension should be part of your daily engine checks. Investigate signs of belt wear, such as black rubber dust shed from the belt or a ragged 'stringy' appearance to its edges.

TOOLS NEEDED

Spanners.

WHERE TO FIND IT

Usually on the front of the engine. In photo A there are separate drive belts for the water pump (lower) and the alternator (upper).

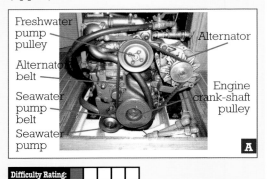

Freshwater pump pulley — Alternator — Alternator belt — Seawater pump belt — Seawater pump — Engine crank-shaft pulley

A

Difficulty Rating:

Adjusting the tension

The method of adjusting the belt tension is pretty crude on the majority of engines. The alternator pivots on a bolt and the swing is restricted by another bolt passing through an adjustment arm. Occasionally a proper screw-threaded adjuster is used.

1 Slacken the nut and bolt on the adjustment arm (photo B).

2 Slacken the main attachment pivot bolt (photo C).

B

C

3 Place a long stout lever between the engine and the alternator so that you can force the alternator outwards to tighten the belt (photo D).

4 While still maintaining pressure, tighten the adjustment and pivot bolts.

Lever the alternator away from the engine to apply tension

Tighten the clamp nut while applying strong force to the lever

D

5 Check the tension and re-adjust as necessary (see below).

6 Adjusting pump belt tension is carried out in a similar manner (photo E).

E

Measuring the tension

You can only measure the tension approximately, unless you have a special gauge. However, a rule of thumb method is sufficient for normal use on belt lengths up to about 450mm.

There are two methods you can use:

THE PUSH METHOD

For short, twin or wide belts:

1 Push the belt at its mid-point.

2 Deflection should be about 10mm (photo F).

Push

F

THE TWIST METHOD

For normal belts:

1 Grip the belt at its mid-point, between the thumb and forefinger.

2 Twist the belt.

Twist

G

3 Tension is correct if it twists approximately 90 degrees (photo G).

Short belt life

If you experience very short belt life, especially if you've fitted a smart regulator or a high power alternator, try fitting a heavy-duty or high-temperature toothed drive belt (photo H).

DAYCO
GOLD LABEL COG-BELT

H

ENGINES: Servicing

9 Servicing the engine

WHY IT NEEDS TO BE DONE

Engines will deteriorate unless they are regularly serviced, and reliability will also suffer. Most leisure boat engines run for relatively few hours each year and actually suffer as a result. An engine that is used infrequently and rarely reaches its proper running temperature is more in need of frequent servicing than one that has harder use.

WHEN IT SHOULD BE DONE

The best guide to servicing is the engine handbook. Although service intervals are normally governed by the number of engine hours since the last service, for the leisure boater it's more often dictated by the need to carry out a service at least annually.

TOOLS NEEDED

A full DIY mechanic's tool kit.

WHERE TO FIND IT

Service points are often scattered about the engine, and seawater and fuel cocks may be found well away from the engine compartment.

Difficulty Rating: ▮▮▯▯▯

(Access may be difficult (or impossible!).)

Cooling system

- Indirectly-cooled engines are cooled, like a car, by fresh water. They need to have the coolant replaced every second year. The coolant should be made up according to the engine handbook's instructions but is often 50% water and 50% antifreeze. Antifreeze contains corrosion inhibitors, which are consumed over time and must be replaced. Checking with an antifreeze tester will not test the inhibitors. Cylinder head gasket failure is common on engines which do not have antifreeze replaced.
- Raw-water-cooled engines (directly-cooled engines) are cooled by sea, lake or river water. They often have sacrificial zinc anodes to prevent electrolytic corrosion of internal parts (photo A). These must be checked annually and replaced if more than half consumed.

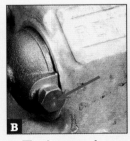
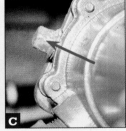

- The heat exchangers of indirectly-cooled engines may have anodes, which need to be checked (photos B and C). These will be found in the raw water section of the engine cooling heat exchanger. Where additional heat exchangers are fitted for oil and gearbox coolers and turbocharger intercoolers, these may also have anodes as well. The handbook should tell you but beware, if the handbook covers several different engine/gearbox combinations only general advice may be given.
- It's generally recommended that the raw water pump impeller is changed annually (photo D).
See *Changing the impeller*, page 12.

- If the seacock is of the 'Blakes' type (Kingston Valve) this will need an annual service – see *Seacocks* chapter, from p108 (photo E).

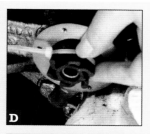

- If your engine is mounted close to the waterline, it will probably have a syphon break. If it has a valve (it will have no

pipe leading to a drain overboard) this must be serviced annually (photo F). Failure to do this can cause water to syphon into the internal working parts of the engine.

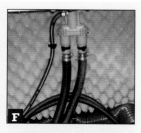

- Check the condition and security of all hose clips (photo G).

Lubrication

- Change the oil 'on schedule' or more often if you have a lot of 'stop/start motoring'.
- Use the grade and viscosity of oil as indicated in the handbook – in the case of older engines the specified grade may not be obtainable so use the nearest. In many

sailing yachts, with minimal engine use, engines never become fully 'run in', even by the time the engine dies a natural death. In these situations, the use of 'synthetic' oil for engine lubrication, unless specified in the handbook, often compounds the problem.

Air system

1 Clean or replace the air filters as indicated on the service schedule (photo H).

2 Check the security of the air filter housing (photo I).

3 If the air filter is connected to the engine with a rubber hose, check softness and delamination.

4 Work on the turbocharger, if fitted, is best left to the experts (photo J).

ENGINE OIL CHANGE

1 Most people recommend that the engine should be run up to warm the oil, making its extraction easier.

2 Then let the engine stand for 10 minutes to allow the oil to settle.

3 Remove the dipstick and insert the tube of the oil extraction pump, trying to get it as far to the bottom as possible (photo K).

4 Pump out the oil (photo L).

5 Pour the required quantity of new oil into the oil filler. (This won't be contaminated by the dirty oil still contained in the old filter, but will give time for the oil to reach the sump.) (Photo M.)

6 Using a suitable wrench, remove the old oil filter, trying to contain the spilled oil (photo N).

7 Fit a new filter, first lubricating the oil seal (photo O). Tighten as indicated on the instructions printed on the filter (photo P).

8 Check the oil level. (As the filter doesn't yet contain any oil it may over-read.) (Photo Q.)

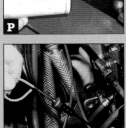

9 Run the engine for a couple of minutes to check for leaks. If you have a mechanical 'stop' control, keep it pulled until the oil pressure light goes out, to allow oil pressure to build before the engine starts.

10 Wait 10 minutes, check the oil level and top up if necessary. If overfilled, remove excess oil.

GEARBOX OIL CHANGE

1 Oil normally has to be removed by a pump, through the dipstick hole (photo A).

2 Refill as specified (photo B). (Note that the gear-box may have its own hand-book.) Sometimes the specification is the same as the oil used in the engine, sometimes it is completely different.

3 For saildrive legs and outdrives the boat will have to be out of the water (except for later model Volvos) to change the oil, because oil is drained from the bottom of the leg (photo C).

Fuel system

1 The fuel tank should be cleaned approximately every five years, but this seldom happens because there's no drain plug and access is difficult. The dirt in this fuel filter bowl indicates the need for tank cleaning (photo D).

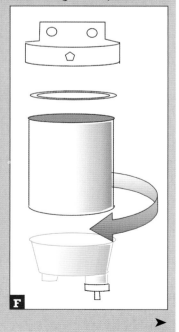

2 Turn off the fuel cock (photo E).

3 Replace the filter element of the pre-filter and clean the water separating bowl (diag. F).

G

4 Clean the filter element (if fitted) of the fuel lift pump (photo G).

5 Replace the element of the engine fine filter (diag. H).

6 Open the fuel cock (photo I).

H

open

I

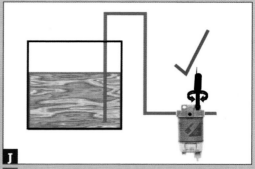

J

7 If fuel level in the tank is above the filter, bleed the filter by opening the bleed screw until fuel runs out then close the bleed screw (diag. J).

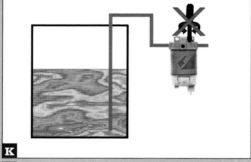

K

8 If fuel level is below the filter do not attempt to bleed (diag. K).

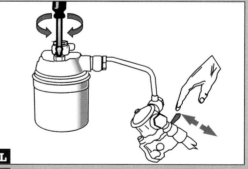

L

9 Bleed the system (diag. L). See *Bleeding air from the fuel system*, page 26.

10 Service or replace the fuel injectors if symptoms dictate – light grey to blue smoke all the time and difficult starting. This needs expert attention.

ENGINES: Other topics

10 Adjusting valve clearance

WHY IT NEEDS TO BE DONE

The engine maintenance schedule will require the valve gear to be adjusted at regular intervals. There is a small clearance between the valve operating gear and the valve stem. Too big a gap will lead to noise or even failure of the mechanism. Too small a gap can lead to the valve not closing properly, resulting in valve failure. Serious engine damage can result from failure to check valve clearances as scheduled by the engine manufacturer.

WHEN IT SHOULD BE DONE

The engine handbook will give details of when this job needs to be undertaken.

TOOLS NEEDED

Spanners, screwdrivers, feeler gauges.

Difficulty Rating:

(Most marine engines have push-rod operated valves and these are simple to adjust. If your engine has twin overhead camshafts, valve clearance is more difficult to adjust and the skill level required is beyond the scope of this book.)

1 If you have a mechanical engine stop system, move the stop lever to 'stop'. Although this is not essential, you will be turning the crankshaft by hand so there is an infinitesimal chance that the engine could start. So if it's easy to do so, you may just as well prevent the engine starting, no matter how unlikely the event (photos A & B). If the lever won't stay in the stopped position, secure it with a piece of cord or wire.

C

2 On this engine (a Yanmar 3 GM30F) there's a breather that needs to be removed first, so that the rocker cover can be lifted clear. Remove the bolts (photo C).

D

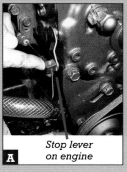
Stop lever on engine
A

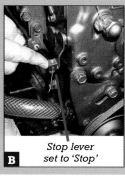
Stop lever set to 'Stop'
B

3 Move the breather out of the way (photo D).

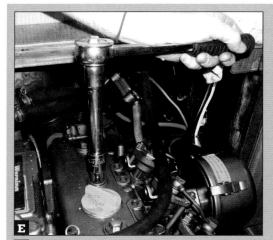

E

4 Undo the bolts holding the rocker box cover in place (photo E).

F1

F2

5 Remove the rocker box cover (photo F1) exposing the valve gear (photo F2).

6 Identify the parts (photo G - ***Names of the Parts***).

Adjuster

Adjuster lock nut

Pushrod

Rocker arm

Gap (valve clearance)

Valve stem

Valve spring

G *Names of the Parts*

7 Workshop manuals normally give instructions of how to position the crankshaft so that the relevant valve is fully closed which is when it should be adjusted. Follow this if you have a manual. However, there is a simple method of ensuring that the valves of a particular cylinder are closed.

8 Identify the inlet valve of one cylinder. This can be done by looking at the inlet and/or exhaust manifolds and visually lining the duct with the valve (photos H and I - ***Identifying the Valves***).

H

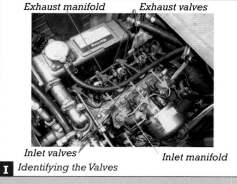

Exhaust manifold   Exhaust valves

Inlet valves   Inlet manifold

I *Identifying the Valves*

J

9 Put a ring spanner or a socket spanner onto the crankshaft nut so that you can rotate the engine (photo J).

10 Check from the engine specification the direction of rotation. Rotate the crankshaft until the inlet valve at the front of the engine opens and closes. This indicates that that cylinder is just starting its compression stroke. Unless both valves are closed air cannot be compressed so this is a positive indication of valve closure. Rotate the crankshaft 180 degrees to 'top of compression' which you can easily feel because the effort required suddenly decreases. Try it a few times so that you are confident you have found 'top of compression'.

K1

K2

11 Select the correct feeler gauge (the clearance is given in the engine manual, which also tells you if the procedure is carried out with the engine hot or cold). (Photos K1 & K2.)

L

12 Slide the feeler gauge into the gap between the inlet valve's adjuster and the pushrod. You should feel a definite drag on the oiled gauge, but you should be able to get the gauge in the gap without difficulty (photo L).

M

13 If the gap needs to be adjusted, slacken the locknut, screw the adjuster in or out to get the correct gap and retighten the locknut. When you retighten the locknut, grip the screwdriver firmly to prevent the adjuster turning with the nut (photo M).

14 Recheck the clearance and re-adjust if necessary.

15 Carry out the same procedure for that cylinder's exhaust valve. Remember that the clearance may be different from the inlet valve.

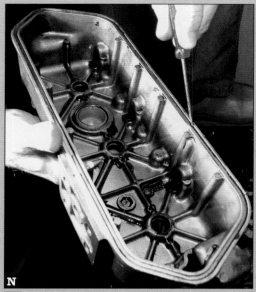

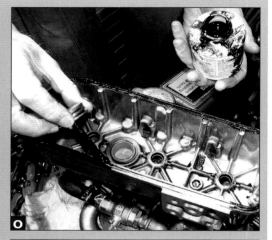

16 Now do the same with the other cylinders.

17 Clean the mating surfaces of the rockerbox cover. Ensure that no dirt gets into the valve gear and the top of the cylinderhead (photo N).

18 Check the condition of the sealing gasket and replace if necessary.

19 Apply a suitable non-setting gasket sealant (photo O).

20 Replace the rockerbox cover (photo P) and any other components that have been removed.

TIP

On Volvo 2000 series engines check that the notch and hole in the gasket are aligned with the breather holes.

11 Engine mounts

WHY IT NEEDS TO BE DONE

Rubber engine mounts will settle as they age. This will cause the engine to become misaligned with the propeller shaft, causing wear within the gearbox. Contamination of the rubber by fuel and oil will cause deterioration of the rubber and possible loss of adhesion between the rubber and the metal parts of the mount. Securing nuts may also loosen, causing misalignment, and

There are two mounts each side

the metal may also fracture, resulting in the engine jumping around. This movement may unseat the stern gland seal resulting in severe seawater leakage.

WHEN IT SHOULD BE DONE

Check the mounts as part of the routine engine service, though it's unlikely to be mentioned in any handbook. Also have a good look at them if you hear any unusual knocking sounds from the engine compartment.

TOOLS NEEDED

None - visual check.

Gap between the rubber and the metal part of the mount

Difficulty Rating: ☐☐☐☐☐

1 Engine mounts need to be checked for settling, which will affect engine alignment and vibration.

- Unless the mount is fully hooded, there is a visible gap between the rubber and the metal of the mount.
- Although it isn't practical to actually measure this gap, you should be able to get your little finger into it (photo A).

A

The weld between the threaded stud and base plate has fractured

B

- If this gap is very small the rubber has settled too much, or even become un-bonded from the metal.

2 Engine mounting brackets need to be checked for fractures.

- Fractures may occur due to ageing (photos B and C).
- Fractures may also occur due to a rope fouling the propeller.

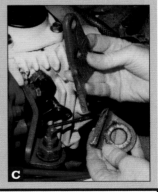

C

TIP An old fracture is often indicated by a black or rust-coloured line at welds or bends in the mount.

12 Engine alignment

WHY IT NEEDS TO BE DONE

If the engine and shaft are not precisely aligned, excessive vibration and gearbox wear will occur. Good initial alignment should have been achieved by the builder. As the engine mounts sag, and as the boat's shape changes under rigging load and by immersion, alignment will be lost.

WHEN IT SHOULD BE DONE

Check engine alignment annually. At the start of each season, correct any misalignment a couple of weeks after relaunch, so that the boat has attained its natural shape afloat.

Most people are daunted by this task, but provided proper initial alignment has been achieved by the builder / engine installer this annual job is not difficult because the sag will cause misalignment only in the vertical plane.

TOOLS NEEDED

Tape measure, feeler gauges, open-ended spanners, hexagonal wrenches, mallet, paint scraper. Possibly a car jack.

Difficulty Rating: ▮▮▮▮□□

Plus patience!

Checking alignment

1. Select forward or astern as appropriate to lock the propshaft.
2. Slacken off the drive coupling bolts by about one turn (photo A).

Loosen coupling flange bolts

A

3. Select neutral.
4. Tap the shaft coupling to make sure it frees from the gearbox output and takes up its natural position. You may need to slacken the bolts some more and use a paint scraper or something similar to

prise the coupling apart (photo B). Next, retighten the bolts until the coupling flanges touch at just one point.

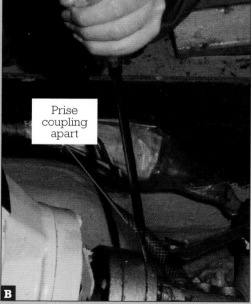

Prise coupling apart

B

5. Make a table like the one on page 42.
6. Rotate the shaft so that one bolt is at the top and mark the coupling so that you know the starting point. Measure the clearance between the gearbox drive

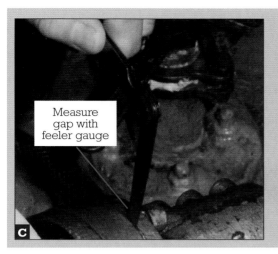

Measure gap with feeler gauge

plate and the coupling plate using a feeler gauge at the 12, 3, 6 and 9 o'clock positions (photo C). Enter in the table.

7 Rotate the shaft 90 degrees. Measure and note the clearances again.

8 Do the same for 180 and 270 degrees, noting the clearances.

9 Add the four figures and divide by 4 for each 'clock position'.

10 For a four inch (100mm) diameter coupling the resulting average should be no greater than 0.004 inches (0.1mm).

11 If all is well, retighten the coupling bolts.

Adjusting the alignment

1 Measure the distances shown on the diagram on page 42 (photo D).

Front mount   Rear mount

2 Enter the distances and coupling clearances into the table on page 42.

3 Measure the number of threads per inch (cm) of the engine mount adjustment bolts to obtain the thread pitch (see diagram, page 42).

4 Enter the thread pitch into the table.

5 Calculate the number of turns for both the forward and aft mounts using the example on page 42.

Loosen lock nut

6 Slacken the mounting lock nuts (photo E).

7 Raise or lower the lower nuts the calculated number of turns (photo F).

Adjust support nut in required direction

8 Tighten the lock nuts (photo G).

Tighten lock nut

Hold support nut securely

9 Recheck the clearance.

ENGINE ALIGNMENT

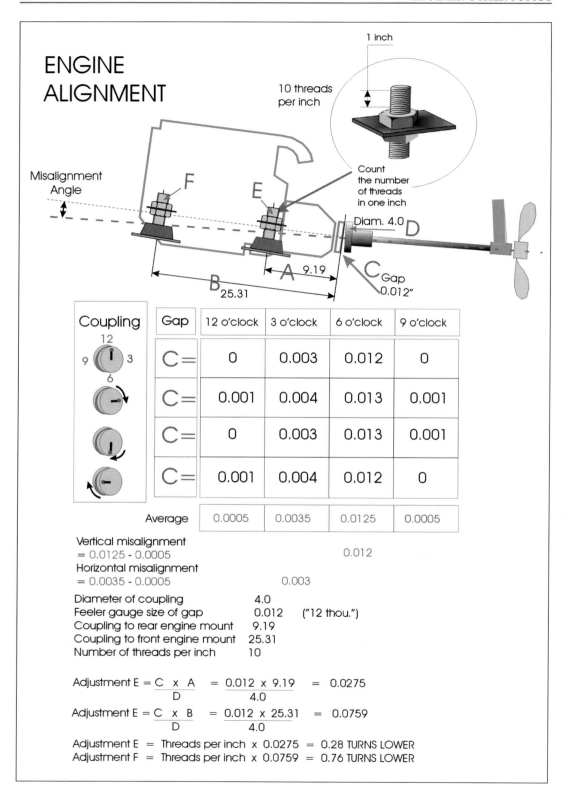

Coupling	Gap	12 o'clock	3 o'clock	6 o'clock	9 o'clock
	C =	0	0.003	0.012	0
	C =	0.001	0.004	0.013	0.001
	C =	0	0.003	0.013	0.001
	C =	0.001	0.004	0.012	0
	Average	0.0005	0.0035	0.0125	0.0005

Vertical misalignment
= 0.0125 - 0.0005   0.012
Horizontal misalignment
= 0.0035 - 0.0005   0.003

Diameter of coupling   4.0
Feeler gauge size of gap   0.012   ("12 thou.")
Coupling to rear engine mount   9.19
Coupling to front engine mount   25.31
Number of threads per inch   10

$$\text{Adjustment E} = \frac{C \times A}{D} = \frac{0.012 \times 9.19}{4.0} = 0.0275$$

$$\text{Adjustment E} = \frac{C \times B}{D} = \frac{0.012 \times 25.31}{4.0} = 0.0759$$

Adjustment E = Threads per inch x 0.0275 = 0.28 TURNS LOWER
Adjustment F = Threads per inch x 0.0759 = 0.76 TURNS LOWER

13 Winterisation

WHY IT NEEDS TO BE DONE

If you are going to lay up your boat for the winter, prepare the engine so that it won't suffer during the period that it's not being used. Given the choice, I would prefer to run the engine under load every couple of weeks, but do winterise the engine if you won't be using it for more than 6 weeks. Any long period in which the engine won't be run requires the engine to receive special treatment and if this period is to exceed 6 months, consult the engine manufacturer or main agent.

WHEN IT SHOULD BE DONE

At the start of the period of lay-up.

TOOLS NEEDED

General DIY mechanic's tools.

WHERE TO FIND IT

You'll be looking at the complete installation.

Checklist
- Make a list of everything done when you lay up the engine.
- Use this list when you recommission the engine to ensure everything is reinstated correctly.

FUEL TANK
- Keep the tank full, especially during the winter, to minimise condensation in the tank.

Carry out a full service of the engine. Then winterise the cooling, air and electrical systems.

Difficulty Rating: ■■□□□

COOLING SYSTEM

1 You will be running the exposed engine. Ensure that neither you nor your clothes become entangled in rotating machinery.

2 If you have direct cooling, run the engine until the thermostat is open. If you don't have a temperature gauge, feel the thermostat housing - when it's hot the thermostat is open (photo A).

3 Stop the engine, close the seacock and disconnect the hose, so that you can dip it into a bucket of fresh water. Restart the engine and when the water reduces to about 25%, add a litre of antifreeze. As the bucket empties, stop the engine. Do not restart the engine until it's recommissioned. This process replaces the seawater with an engine-friendly liquid (photo B).

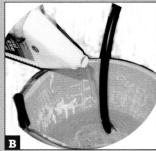

4 Don't remove the water pump impeller. Change the impeller when you recommission, or you may lose some of the water/antifreeze mix.

5 If you have an indirect cooling system, it's not a bad idea to flush the seawater part as above, but the thermostat doesn't have to be open – because it's in the fresh water part of the system. Seawater remaining in the system could freeze in very low temperatures and draining the system isn't always fully effective. If there's a gearbox and engine oil cooler, these will have seawater in them and will now be protected by the antifreeze.

AIR SYSTEM

1 Disconnect the exhaust hose from the engine and check its condition.

2 Remove air cleaner(s) from the engine.

3 Spray water-repellent oil up the exhaust and air intakes and then seal with plastic bags to prevent damp air entering (photo C).

4 If the boat is to remain afloat, block the exhaust where it leaves the hull or reconnect the exhaust to prevent water flooding the boat. Otherwise leave the exhaust open to ventilate the hull.

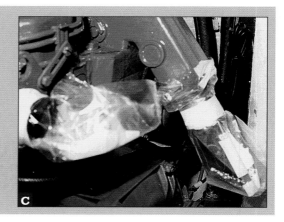
C

ELECTRICS

1 Check the battery levels (photo D).

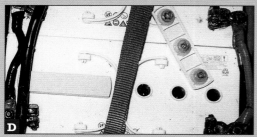
D

E

2 Top up the batteries with distilled water if necessary (photo E).

3 The batteries need to be recharged monthly so take them home if necessary.

4 Clean and tighten all electrical connections - if in doubt, renew them - and remember the negative cables on the engine block.

CLEANLINESS

1 Clean the engine and engine compartment.

2 Remove any rust, then touch up the paint.

3 Spray the engine with a water displacing fluid such as Boeshield T 9 (photo F), including the wiring and terminals.

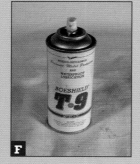
F

Ventilation and heating

To help prevent condensation:

1 Open all access panels to the engine compartment to allow ventilation.

2 If mains power is available:

- Use a low-powered tubular heater in the engine compartment.
- Run a dehumidifier. Although relatively expensive to buy, weekly running costs are not great and you will be protecting your investment.

14 Engine won't start

ENGINE WON'T TURN

1 Is battery 'on'?

2 Is panel alive? - Warning lights 'on'?

3 Check engine fuse – (usually on engine) - but there may not be one. Small Yanmars have one but it isn't mentioned in the handbook - it's 30 amp and in the harness, close to the starter motor. It's taped over and sprayed with engine paint (photo A).

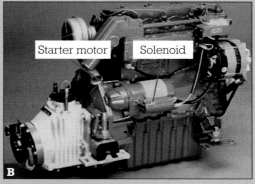

Starter motor   Solenoid

B

Fuse holder

A

4 If the warning lights go out when you operate the starter, the battery is flat or the connections are very poor.

5 If the starter solenoid is not receiving current you may be able to 'jump start' the engine. The instruments and warning lights may not then operate (photo B).

6 If it is receiving current you should be able to hear the solenoid click - if you can't then the contacts have failed or the engine/starter has seized.

7 Older engines have a spiral groove along which the gear moves to engage the flywheel. This may need cleaning. If the starter just 'whirs' the starter pinion is sticking on its spiral thread. If it's stuck at the flywheel end, and the starter won't turn, you can 'wind' it back to the starting position using a spanner on the front end of the starter shaft.

JUMP STARTING

1 Make sure that neither your hair nor anything you are wearing can get caught up in the machinery.

2 Bridge the positive (battery) terminal on the starter solenoid and the starter switch terminal on the solenoid to turn the engine (diagram C).

3 This won't work if the solenoid has failed.

TIP On my boat, I have a switch permanently wired into the jump-starting bridge. This switch is installed in a safe position in the engine compartment so that it can be used without risk. It's very useful for servicing the engine, too.

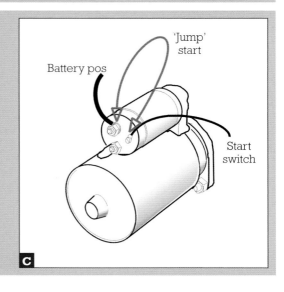

'Jump' start

Battery pos

Start switch

C

ENGINE TURNS BUT WON'T START

Turn off the cooling water seacock in a sailing boat before continuing – turn on again when engine starts. Don't run the starter motor for more than 30 seconds without an intervening 5 minute cooling period.

Engine turns slowly:

1 Low battery state - if possible, use batteries in parallel.

2 Check battery connections – remember the battery connection to the engine block.

3 If you have decompressers, decompress the cylinders while you turn the engine on the battery so that full turning speed is achieved, then close them (or only one if the system allows) to achieve a start (photo D).

Engine turns normally:

1 Check 'stop control' is not at 'stop' (diag. E).

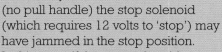

2 If the engine is stopped electrically (no pull handle) the stop solenoid (which requires 12 volts to 'stop') may have jammed in the stop position.

3 In this case, if the stop solenoid operates externally, move its lever by hand (photo F).

4 Or, if the mechanism is internal, remove the stop solenoid and operate the stop switch to see if the operating rod moves.

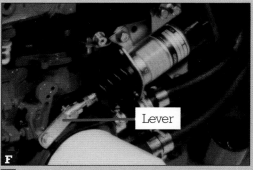

Lever

5 Some marinised engines (and generating sets) need 12 volts to run so make sure the connections are sound.

6 Check fuel 'on'.

7 Check fuel contents.

8 Check 'cold starting procedure' is correct – read the handbook.

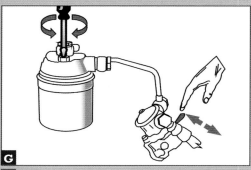

9 Bleed low pressure fuel system. Make sure it's fuel not water (diag. G).

10 Bleed high pressure fuel system (photos H and I).

Bleed the high pressure line either at the injection pump…

…OR, at the injectors, whichever is more convenient

DO NOT USE STARTING FLUID ON A DIESEL ENGINE - DAMAGE IS LIKELY

No current diesel engine manufacturer approves the use of starting fluid.
(If you must, spray some on a rag and hold the rag by the air intake as the engine is cranked. Get the engine overhauled!)
A much safer way is to apply heat to the intake area and preheat the air as it's sucked into the air intake. Remove the air cleaner and apply heat with a hair dryer.

In **DIRE EMERGENCY**, using exceptional care to prevent fire and burning electrical circuits, a gas blow-lamp can also be used.

Remember that difficult starting from cold indicates a faulty technique, poor battery or connections, or an engine problem - which can include a bent connecting rod caused by water in the cylinder.

CHECKLIST - ENGINE WON'T START

Does starter turn engine?
- No

1. Battery switch - is it on?
2. Battery - is it flat?
   Watch voltmeter as you start.
3. Fuse - is it blown?
   Do you know where it is?
4. Starter switch - has it failed?
   Are connections sound?
5. Solenoid - has it failed?
   Check connections - can you hear it 'clunk'?
6. Starter motor - has it failed?
   Check connections.
7. Seized engine -
   see your bank manager!

Does starter turn engine?
- Yes

1. Fuel - is it on?
   Is there enough?
2. Cold start device - do you need it?
   Have you used it correctly?
3. Compression - is there sufficient to heat the air to ignition temperature?
4. Is engine turning fast enough?
   Check battery volts.
5. Is there enough air?
   Is there a blockage?

15 Engine won't stop

ENGINE WON'T STOP

Diesel engines are stopped by cutting off the fuel supply at the injection pump. Prepare for the day it fails - familiarise yourself with how it works by getting someone to operate the stop control. Look at the engine in the region of the fuel injection pump to see what moves (or makes a noise in the case of an enclosed solenoid mechanism).

If it won't stop the mechanism has failed.

MECHANICAL STOP CONTROL

Operate the stop lever on the engine (diags. A and B).

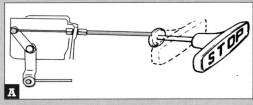

ELECTRICAL STOP CONTROL

The electrically operated stop control needs 12 volts to stop the engine (photo C).

1 If the connection has failed, connect its electrical terminal to a 12 volt supply.

2 If the solenoid has failed, you should be able to operate the lever manually.

3 If the solenoid is mounted directly on the engine, removing the solenoid should shut the engine down (photos D and E).

4 Some marinised engines (and generating sets) need 12 volts to run, so in all probability the solenoid has seized. **Disconnect or remove the solenoid.**

FUEL SYSTEM

16 Fuel contamination

WHY IT NEEDS TO BE DONE

Contaminated fuel will stop the engine.

Initially, the engine will not run at high rpm, and a diesel may well hunt as the governor tries to accelerate the engine but lack of fuel prevents this. Gradually the engine will get slower and eventually refuse to start. Petrol engines running on too little fuel will initially accelerate as the fuel/air mixture weakens, but then die.

Bleeding the diesel fuel system will reveal that no fuel is flowing from the tank. Undoing a fuel union to check fuel flow on a petrol engine in a boat will need great care to prevent petrol dripping onto a hot component and petrol vapour accumulating: an explosion waiting to happen.

WHEN IT SHOULD BE DONE

A daily check of the water separating fuel filter should prevent fuel starvation (other than mechanical failure), but the outlet from the tank could become blocked without any tell-tale debris in the filter bowl.

Difficulty Rating: ▮▮ ▯▯▯▯

(But skill and patience are needed if you have to remove the tank.)

TO PREVENT CONTAMINATED FUEL REACHING THE ENGINE

Filter

A

A water-separating primary filter should be fitted to prevent contamination reaching the engine (photo A). Long distance cruisers may choose to fit twin, parallel, primary fuel filters (photo B).

B

THE CONTAMINANTS
1. Water

C

- Ensure that your tank filler seal is in good condition (photos C & D).
- Keep a spare seal.
- Keep your tank full, especially in winter, to reduce condensation.

Seal

D

2. Dirt

- Ideally, clean your tank every 5 years.
- This is often difficult with many boats, if not impossible.
- If dirt and rust debris accumulate in the fuel tank, the tank's outlet will eventually become blocked.

3. Bacteria

- With any water present in a fuel tank, algae and fungal spores will multiply at the interface between the fuel and the water and this is when they become a problem (photo E).

Sludge

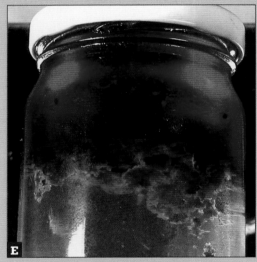

E

- Algae and fungal growths will cause sludge in the tank.
- The sludge may be acidic, causing corrosion of all components of the fuel system.
- The sludge can rapidly multiply, causing blockage of the fuel system (photo F).

THE SOLUTIONS
A. Fuel conditioners

Various additives are available to give improved combustion, flow and corrosion inhibition; you only have to read the labels to see what they are claimed to do. Unfortunately you have to take these claims on trust as there is little real evidence supplied. However, it is probably true to say that some improvement is likely when used with marine diesel, which is not well endowed with additives, and you'll probably find several types in your chandlery.

Magnetic fuel conditioners are available with similar claims. I have never managed to get any scientific proof of their effectiveness and Which? magazine, a number of years ago, tested one which met none of the claims.

B. Water eliminators

- Additives are available that claim to take up any water in the fuel, passing this water through the system in microscopic size, and injecting it harmlessly into the combustion chambers. They are used in the road transport industry to good effect. Have a look to see which your chandler stocks.
- However, there's a big difference between road transport and marine leisure use. HGVs run their engines for a large portion of the day and are not idle for long. Leisure marine craft rarely run their engines at all.

 TIP Don't use a biocide at each refuelling.

 TIP Daily inspection of the water separating fuel filter will alert you to problems.

This I believe leads to two basic problems:

- Fuel with entrapped water will be sitting in the fuel injection pump for weeks at a time with the probability that corrosion will occur inside the pump.
- The fuel/water interface area available for the growth of algae and fungal organisms will be hugely greater with the water in suspension than with the water at the bottom of the tank.

C. Biocides

- Biocides kill bacterial organisms.
- Continuous use of a biocide (as suggested in the instructions) could produce a breed of 'super bug' immune to the biocide.
- The chemists who design the biocides say that they should be used only as required and not as a preventive.
- It would probably also be a good idea to use a different one for each 'infestation'.
- You will usually find several makes of biocide on your chandler's shelves.

D. Other additives

- 'Soltron' is an enzyme additive that came to my attention several years ago and I have full lab reports of its effectiveness (photo G).
- This will be found in some chandlers, but if you don't see it visit www.soltron.co.uk
- Basically, it conditions fuel to increase combustion efficiency. It removes sludge (but not scale and dirt) from the tank and converts it back into fuel and it kills 99%+ of bacteria.
- As an enzyme, it may be used safely at each refuelling, and repeated use ensures that all bacteria are killed.

17 Cleaning the fuel tank

Tank cleaning may be very difficult for boat owners who find that they have no drain sump and are unable to drain their tanks or clean them – and cannot remove them either!

Photo A shows a drain sump.

Some methods that you may be able to use to attempt to clean your tank:

1 Drain the tank using a flexible pipe pushed into the fuel filler pipe to reach the bottom of the tank, connected to a 'fuel safe' pump. This is unlikely to remove sludge and much in the way of dirt and debris. A Pela vacuum pump may do the job better (photo B).

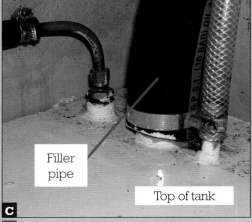

Filler pipe

Top of tank

2 Remove the tank filler pipe from the top of your tank if you can get at it (photo C). You may then be able to 'wriggle' the suction pipe around a bit.

3 Cut an inspection hole into the top of the tank if you can get at it. This will allow access for cleaning as well as draining. Make a new hatch to seal the hole and bolt it in place with a gasket.

4 Cut an inspection hatch in the structure of the boat to allow tank access.

5 **DO NOT ATTEMPT TO DRILL OR CUT A PETROL TANK, EVEN IF YOU BELIEVE IT TO BE EMPTY.**
An explosion is likely.

WATER SYSTEM

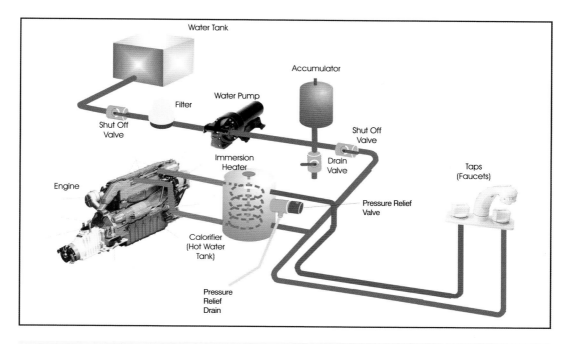

Water Tank

Accumulator

Water Pump

Filter

Shut Off Valve

Shut Off Valve

Immersion Heater

Drain Valve

Engine

Taps (Faucets)

Pressure Relief Valve

Calorifier (Hot Water Tank)

Pressure Relief Drain

A basic water system consists of no more than a water tank supplying a manually operated pump and a tap.

Many boats have a more complex system.
- An electric pump will supply water from the tank to an on/off tap. To stop pressure surges that can damage the pump, an accumulator may be fitted. This is a pressure vessel that stores water under pressure, allowing better control of the water flow. The pump switches on only when some of the pressure has dissipated.
- Heat can be stored in a calorifier so that domestic hot water is on tap. The heat may be from the engine, a mains electric immersion heater, or a gas or diesel-fired boiler that is part of the boat's heating system.

Water system
In the sections below, bear the following in mind.

- Servicing the water system often entails isolating the water pump, filter and accumulator. It can be a great help to install isolating valves and drain valves.
- Fresh water in the bilge may be from the hot water tank pressure relief valve.
- The pump filter will need regular cleaning.
- If there's an accumulator that does not have a diaphragm, the air in it will gradually be absorbed by the water so the air will occasionally require replacement.
- If you have a water taste filter or, even better, a purification filter, the element will need regular replacement.
- Contaminated pipes may need to be replaced.
- Tap washers may require occasional replacement.
- The water pump may require an overhaul.
- Pipe fittings may need changing.

18 Recharging the accumulator

An accumulator is a pressure reservoir that stores pressurised water so that the pump does not have to run as soon as you open a tap. This leads to longer pump life and quieter operation.

THE PROBLEM

If the pump runs as soon as you turn the tap on, either there is no accumulator or, if there is one, it's lost all its air.

Inflation valve

There are two types of accumulator:

- The cheaper type has no diaphragm to separate the water and the air. Initially the accumulator contains only air. Water, pressurised by the pump, is forced into the accumulator, compressing the air, thus providing a reservoir of water under pressure ready for instant use and to smooth out pressure fluctuations. Over time, the compressed air is absorbed by the water and needs to be recharged. (Photo A.)

- More expensive accumulators use a rubber diaphragm to separate the compressed air from the water, the air being compressed prior to use by a bicycle pump. Rubber is slightly permeable and over time the air pressure will reduce and will need to be topped up by a bicycle pump. (Photo B.)

TOOLS REQUIRED

A screwdriver.

Difficulty Rating: ☐☐☐☐☐

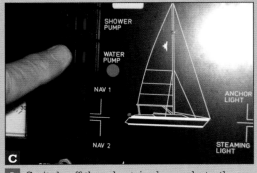

1 Switch off the electrical supply to the domestic water pump (photo C).

2 Shut off the water supply to the pump using woodworking clamps or isolating valves (photos D and E).

F

G

3 Open the air valve on top of the accumulator (photos F and G).

4 Open the line drain valve, if there is one. If not, undo the pipe clamp (Jubilee clip) and drain the accumulator of water.

5 Close the drain valve.

Crack

H

6 Close the air valve. (If you tighten this too hard, the neck of the accumulator will crack) (photo H).

7 The accumulator is now full of air and empty of water.

8 Replace the pipe (if removed), or close the drain valve.

9 Open both isolating valves or remove the clamps.

I

10 Open the water tap furthest from the pump (photo I).

11 Switch on the electrical supply to the water pump.

12 When the water is flowing smoothly, turn off the water tap.

13 The pump should continue to run while it is pressurising the accumulator and then stop automatically.

J

14 Check the accumulator (and any pipes you disconnected) for leaks (photo J).

 TIP Consider installing a drain cock for servicing the accumulator.

19 Replacing contaminated pipes

The water system's filler and vent pipes are open to bacterial and fungal contamination because they are open to the air. Even if the tanks are occasionally dosed with a sterilising agent such as chlorine, this probably won't reach the filler and vent pipes. In many boats they are transparent, so their condition can be seen easily.

TOOLS REQUIRED
Screwdriver and hobby knife.

Difficulty Rating: ▮▮□□□

1 Locate and undo the pipe clips (Jubilee clips) at each end of the pipe.
2 Remove the pipe.
3 Cut new length of pipe.
4 Install new pipe.
5 Tighten pipe clips.

20 Hot water pressure relief valve

Owners often find unexplained fresh water in the bilge. Each time the hot water is reheated it will expand, and the pressure exerted may cause the pressure relief valve on the hot water tank (calorifier) to operate. This water will discharge into the bilge and is a sign that the pressure relief valve is working properly. If the water in the tank were always hot, the valve would operate only once, on initial warm up.

On a boat, this may happen frequently as the tank is alternately heated and cooled. The water released by the pressure relief valve can be collected in a bottle and drained when necessary.

TOOLS REQUIRED
Screwdriver and hobby knife.

Difficulty Rating: ▮□□□□

Collecting water from the hot water pressure relief valve

A   B   C   D

1 Fit a length of plastic hose onto the outlet of the pressure relief valve and lead it to somewhere accessible (photos A and B).
2 Allow this pipe to drain into a plastic bottle (photos C and D).
3 Inspect the bottle and drain as necessary.

21 Replacing the drinking water filter

THE PROBLEM
The filter will need regular replacement, either because the water flow has reduced due to filter clogging, or because it has reached the time limit beyond which bacterial growth inside the filter becomes a distinct possibility (normally 6 months).

TOOLS REQUIRED
Screwdriver.

Difficulty Rating: ☐ ☐ ☐ ☐

1 Turn off the electrical supply to the domestic water pump.

A

2 A simple disposable taste filter is generally held in place by two pipe clips (photo A).

B

3 A bacterial filter has a replaceable element (photo B).

C

4 Remove the filter bowl, observing the manufacturer's instructions (photos C, D and E).

D

E

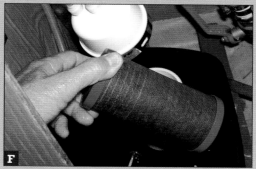

F

5 Remove the old filter (photo F).
6 Insert the new filter.
7 Replace the filter bowl.
8 Restore the electrical supply to the pump.
9 Turn on the drinking water pump and let the water run until it becomes clear.

TIP As you will inevitably spill some water as you do this job, unhook the filter body from its mounting (if possible) and do the job over a container small enough to go under the filter.

22 Installing isolation and drain valves

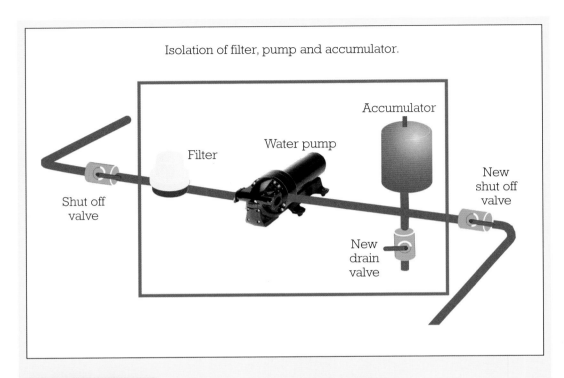

Isolation of filter, pump and accumulator.

Accumulator

Water pump

Filter

New shut off valve

Shut off valve

New drain valve

TOOLS NEEDED
Pipe cutter, clamps, screwdriver.

Difficulty Rating:

Installing isolation and drain valves

1 Switch off the water pump (photo A).

2 Clamp the water pipes to isolate the length of pipe that you are going to cut (photo B).

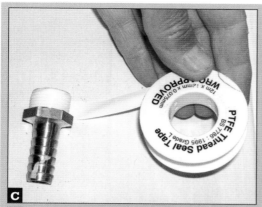

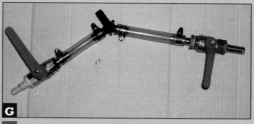

3 Apply PTFE pipe joint tape to the thread of the hose tails (photo C).

4 Fit the hose tails to the stopcock (photos D, E and F).

5 'T' an isolating valve and a drain valve into the pipework downstream of the accumulator (or pump if no accumulator is fitted) (photo G).

6 Fit a shut off valve upstream of the filter (photo H).

7 Remove the clamps.

8 The filter, pump and accumulator can now be easily drained using a pipe attached to the drain valve (photo I).

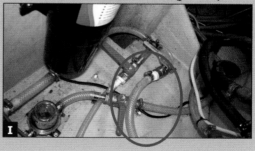

23 Leaking tap washer

THE PROBLEM
The tap drips water. This is much less likely on a boat system than at home, as the water pressure is lower and the tap is used less.

TOOLS REQUIRED
Open ended spanners, screwdrivers.

Difficulty Rating:

1 Switch off the water pump.

2 Insert the plug into the sink drain to prevent dropped articles going down the waste pipe (photo A).

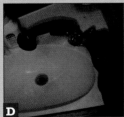

3 Remove the tap handle. You may simply have to pull it off, or you may have to remove the top central cover to expose the securing device (photos B, C, D, E, and F).

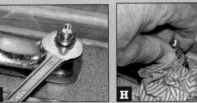

4 Unscrew the tap body (photos G, H and I).

5 Remove the old washer. This may be a push-on type or you may have to remove a retaining screw (photos J and K).

6 Replace with a new washer.

7 Reassemble the tap.

TIP If you haven't got a new washer, turning the old one over to use the other face often suffices (photo L).

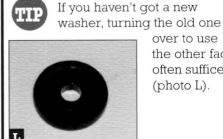

24 Plastic water pipe fittings

THE PROBLEM

Modern boats have semi-flexible opaque water pipes and fittings as are often found in domestic water systems. They have easy push-fitting components, without soldered joints. Any work you do on the pipe runs, including installing a new component, needs knowledge of how the fittings work.

TOOLS REQUIRED

Pipe cutter, hobby knife.

Difficulty Rating:

1 Cut the pipe with a pipe cutter, not a saw (photo A).

2 This may not cut the pipe fully due to its flexibility, so you may have to finish the job with a hobby knife (photo B).

3 The fittings differ a little according to make, but a typical joint is shown in photos C and D.

4 Some joints can be reused, while others can't be remade but are cheaper. For boat use it's probably worth paying the extra for those that can be reused.

5 The pipe insert is essential to stop the plastic pipe being crushed (photo E).

6 The fittings can be used with copper pipe, and in this case the pipe insert is not required (photo F).

7 Different pipe diameters and pipe materials can be accommodated (photo G).

25 Water pump

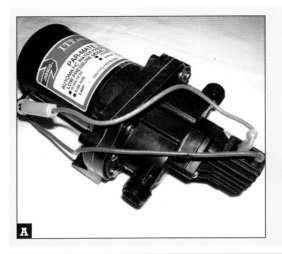

A

THE PROBLEM

The domestic water pump fails to run, or runs but fails to pump water. This may indicate a blown fuse, a failure of the pressure switch, failure of the pump diaphragm or valves, or a failure of the electric motor. If the fuse is found to be OK, then a troubleshooting routine is needed.

TOOLS REQUIRED

Depends on pump – screwdrivers as a minimum.

Difficulty Rating: ■■■□□□

If the pump runs but no water flows

Check the water supply.
1. It's probable that there has been a failure in the water pump mechanism.
2. A spares kit may be obtainable, in which case full instructions will be found in the kit.

B

3. You will have to remove all the pump body screws (photo B).
4. Take care not to lose any parts.

If the pump does not run

1. Check the fuse. As well as one on the panel, there may well be an in-line fuse in the supply wiring at the pump (photo C).

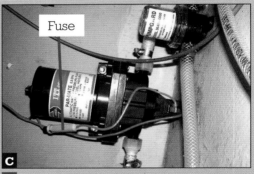

Fuse

C

2. If pump still does not run with at least one tap open, or fails to pump: remove the wires from the pressure switch (photo D) and join them together.

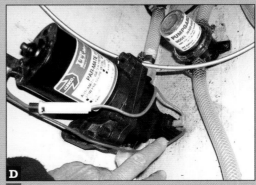

D

3. If the pump runs, the pressure switch is at fault.

Checking the pressure switch

1 It may be possible to check the pressure switch with the pump still mounted in position. If so: remove the pressure switch cover (photos E, F and G).

E

F

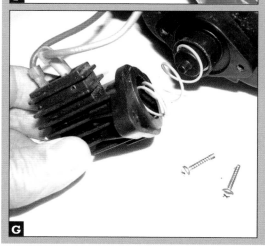

G

H

2 Push the micro switch button (photo H).
3 If the pump runs, the electrical switch is OK.

I

4 Push the pressure plunger to ensure that it moves in and out (photo I) – it moves only about 2 or 3mm but it should move. If it doesn't, it's seized and so can't operate the electrical switch - you will have to take the pump apart.

If you can't remove the cover with the pump in place, you will have to remove the pump:

5 Turn off the power to the pump.
6 Turn off the water supply, using a clamp if necessary.
7 Disconnect the wiring, undo the pipework, remove the mounting bolts/screws and remove the pump.
8 Reconnect the wires and switch the power back on.
9 Carry out steps 1 to 4 above. If none of the steps produces results, you will have to replace the pump.

26 Water pump filter

A

THE PROBLEM
The water pump should be protected from debris by an inline water filter. This normally has a transparent bowl. If it appears that the filter is dirty, it will need cleaning (photo A).

TOOLS REQUIRED
Probably none, but you may need a woodworking clamp if there's no isolating stopcock.

Difficulty Rating: ▰▰▱▱▱

1 Shut the filter's isolating valves.

B

2 If there are no valves, use woodworking clamps to pinch the flexible pipe supplying the filter (photo B).

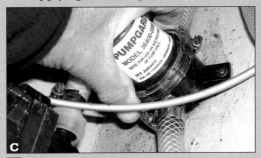

C

D

E

F

3 Unscrew the filter bowl (photo C).
4 Remove the filter element (photos D and E).
5 Clean the element under running water.
6 Replace the element (photo F).
7 Replace the bowl, ensuring that the sealing ring is installed correctly.
8 Turn the water supply back on, and check for leaks.

 TIP If there's no water stopcock, fit one now, to make future maintenance easier.

GAS SYSTEM

Special regulations of the Recreational Craft Directive (RCD) and the Boat Safety Scheme cover boats in various categories and it's essential that these regulations be observed. In the UK, competent engineers should carry out work on a gas system, so it's recommended that a CORGI registered engineer do this work.

Guidance on the Boat Safety Scheme is available from British Waterways or National Rivers Authority.
Guidance on the RCD is available from www.rya.org.uk/infoadvice/regssafety/Pages/recreationalcraftdirective.aspx
Guidance on LPG installations is available from Calor Gas on 0800 626 626 or www.socal.co.uk

Butane (blue bottle) is normally used in

Europe, but it will not work in cooler temperatures, in which case Propane (red bottle) is used. Butane and Propane run at different pressures and have different regulators. Calor recommend that if you do need Propane, it's sensible to use this all year to avoid having to change regulators.

Gas is stored in the bottle as a liquid under pressure. You can feel it sloshing around if you shake the bottle! It isn't filled completely with 'liquid gas', so there's a layer of 'gaseous gas' at the top of the bottle. A gas pressure regulator reduces the pressure of the gas supplied to the appliances to a safe, useable, value. The gas bottle must always be used 'upright' or the regulator will be filled with liquid gas, causing serious problems.

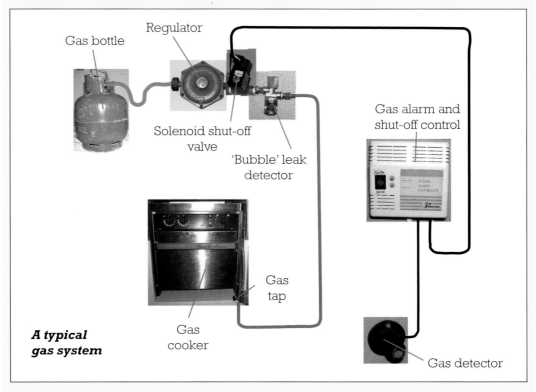

A typical gas system

Gas bottle

Regulator

Solenoid shut-off valve

'Bubble' leak detector

Gas alarm and shut-off control

Gas cooker

Gas tap

Gas detector

27 Checking for leaks

- The number of joints in the pipework must be kept to a minimum.
- Check these joints for leaks on a regular basis.

WHY IT NEEDS TO BE DONE
LPG (liquid petroleum gas) is heavier than air and will accumulate in the boat's bilge and form an explosive mixture of gas and air.

WHEN IT SHOULD BE DONE
At least annually, and at any time you suspect a gas leak.

TOOLS NEEDED
Small paintbrush and some liquid soap solution.

Difficulty Rating:

1. Put some washing up liquid and water into a small jar.
2. Paint all joints with the solution
3. A leak will show up as bubbles forming at the joint and increasing in size.

TIP Fit a 'bubble' leak detector (photo A). This is installed between the gas bottle and the regulator. If a leak is present anywhere in the system, you will see bubbles in the sight glass when you press the red knob. It's important to follow rigorously the instructions for testing supplied with your unit.

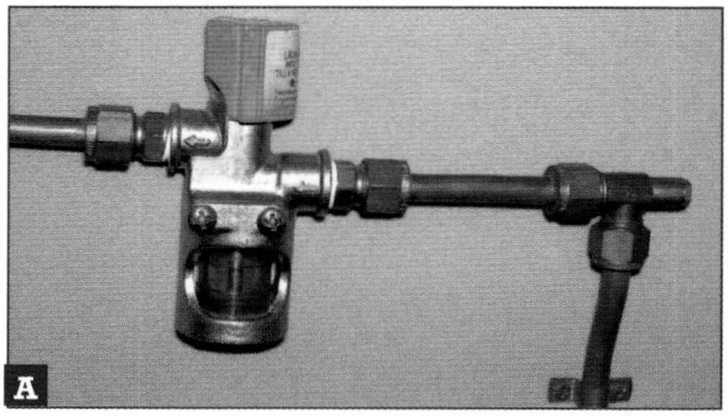

A

Replacement of gas components

WHY IT NEEDS TO BE DONE
Certain components of the gas system have a restricted life and will need to be replaced at intervals.

WHEN IT SHOULD BE DONE
- Regulators: every 10 years.
- Flexible hoses: every 5 years, or earlier if there are signs of leaks, flaws, brittleness, cracking, abrasions, kinking, or significant bleaching.

28 Do I need to replace the gas system components?

Checking the age of the regulator

B

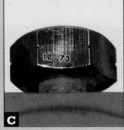

C

1 On this type (photo B), look on the nut (photo C). This one shows the date of manufacture as December 1973.

D

2 On this one look on the back. The numbers show 1998 and the two dots represent February (photo D).

E

3 If the regulator looks like this, change it anyway (photo E).

Checking the age of flexible hose

1 If it's bleached or perished, change it (photo F).

F

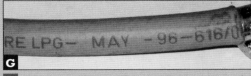

G

2 If it's out of date, change it. The date of manufacture of this hose is May 1996 (photo G).

Check the condition of the components

H

1 Is the gas pipe corroded? This pipe shows signs of dezincification (photo H).

I

2 Broken armour protection (photo I).

J

3 Corroded heater exhaust (photo J).

4 Kinked gas pipes will be weakened and will reduce gas flow (photo K).

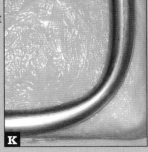

K

29 The ideal gas system

- Has a solenoid shut-off valve connected to the regulator.
- Has a bubble leak detector close to the cut-off valve.

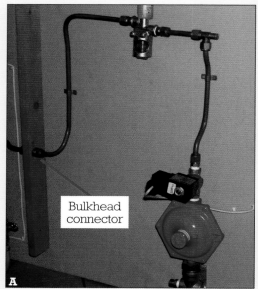

Bulkhead connector

A

- Uses bulkhead connectors where the gas pipe passes through a bulkhead (photo A).
  Preferably (to minimise joints) is run in a gas-tight conduit (transparent to allow visual inspection).

B

- Has a shut-off tap prior to each appliance (photo B).

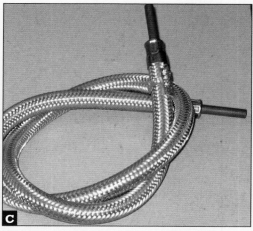

C

- Uses armoured hose for a swinging cooker (photo C).

D

- Has a gas alarm (photo D), which sounds if gas is detected by the detector (photo E), AND shuts the gas solenoid valve.

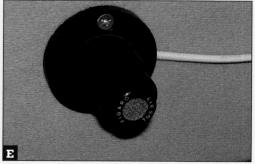

E

IF YOU DO HAVE TO DISCONNECT A GAS PIPE

PFTE

- Use PTFE tape (photo A) or an approved gas sealing compound such as Calortight (photo B) to join threaded components.

- Joining gas pipe to components with 'olives' needs no sealing compound. Photo C is a cutaway showing how it should be done.

- It doesn't need to be plastered with Calortite (photo D).

- Don't overtighten compression joints (photo E).

- The olive, after compression, should look like this (photo F).

ELECTRICAL SYSTEM

30 Formulae

BASIC ELECTRICAL KNOWLEDGE

1 Resistance of a wire or component is measured in Ohms (Ω).

2 Volts (V) drives the current through the wire.

3 Current (I) is the flow of electricity through the wire. It's measured in Amps (A). (Strictly, current is in amperes but most people use the word amps.)

4 Power is the voltage multiplied by the current: Power = VI Watts.

5 Ohm's Law tells us that the current through a wire is calculated by dividing the volts by the resistance – thus: $I = V/R$ and so $V = I \times R$, $R = V/I$.

6 Voltage loss in a long wire run should not exceed 3%. On many boats the loss is as much as 10%. This gives dim lights and wastes power.

7 The resistance of several components connected 'in series' is the sum of their individual resistances. The same current flows through all of them. The system voltage acts over the complete string of components. ($R = R1 + R2 + R3$ etc.)

8 The resistance of components connected in 'parallel' is a little more complex and is found by:

$$R = \frac{1}{1/R_1 + 1/R_2 + 1/R_3 \text{ etc.}}$$

- For only two resistances this is simplified to:
$$R = \frac{R_1 \times R_2}{R_1 + R_2}$$

Energy consumption from a battery

- The energy consumed from a battery by a piece of equipment is measured in amp hours (AH), the length of time it's switched on multiplied by the current flowing through it.

- A 25 watt navigation light has a current of 2.08 amps at 12 volts (power / volts). Switched on for 8 hours it will consume 16.64 amp hours (amps x time).

Note: Many authors erroneously state this consumption in amps with no reference to the time the current is flowing.

Electronics

Electronic equipment will normally run on a large voltage range: 9 –16 volts or even 9 – 32 volts.

Heavy load items

Fridges, heaters and other heavy load equipment will often be set to cut out if the voltage drops below a critical level.

Note: The voltage drop across a fuse protecting a heavy load may cause the load to cut out. Heavy load items should be protected by a circuit breaker rather than a fuse, to prevent this.

31 Multimeter

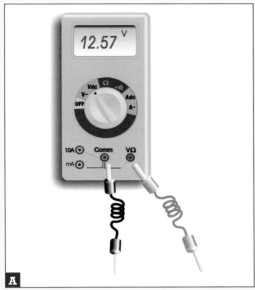

A

B

Troubleshooting and maintenance of the electrical system is enhanced by the use of a multimeter (diag. A). These meters can be purchased for a modest price from electronics stores and for general use an auto-range meter is probably most appropriate, although a manual meter is cheaper. With manual multimeters you need to estimate what the value is before you test it.

A small probe-type multimeter (photo B) frees up the hands, allowing you to test the circuit and read the meter simultaneously.

Multimeters have an internal battery and so must be switched off when not in use.

SERIES & PARALLEL
(drawing C)

- Two batteries connected together in line (in SERIES) have their voltages added.
- Two batteries connected together +ve to +ve and –ve to –ve (in PARALLEL) have their capacities added but the voltage remains the same as a single battery.
- Bulbs or other equipment connected to the same supply must be connected in PARALLEL.

(Battery capacity is the energy it contains in amp hours when fully charged.)

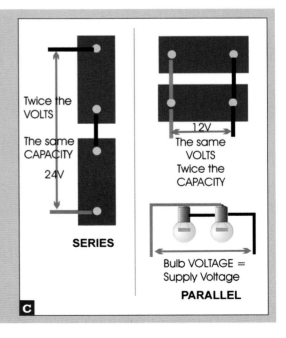

C

Checking continuity
(drawing D)

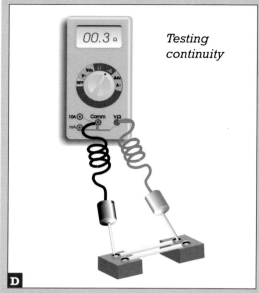

Testing continuity

D

Check continuity of, for example, a wire by checking its resistance. Low resistance = OK, High (infinite) resistance = a break.

1 Set the meter to resistance (Ω) (and a low range for a manual meter).

E

2 Hold the probes together in contact to check for zero reading and 'beep'. Analogue meters need to be 'zeroed' at this stage – if it can't be zeroed the internal battery is low (photo E).

3 Switch the circuit to be tested to 'OFF'.

4 Put the probes at each end of the wire to be tested. The resistance should be zero but will probably read several ohms because of the resistance of the wire.

5 If set to 'bleep' you will get an audio warning of very low resistance for a continuity check. A 'blinking' reading indicates an open circuit (i.e. a break).

6 If the length of circuit is longer than the probe wire, use a long length of wire to extend the probe. (Keep a long length of 1.5mm² wire specially for the purpose.) This allows a long single conductor to be checked.

7 Measurement of the resistance of a component can be made only with the component isolated (otherwise the rest of the circuit may influence the reading).

Testing a bulb
(drawing F)

F

1 Put the meter's probes onto the bulb's contacts and measure its resistance.

• If there's only a central contact, the metal casing of the bulb is the other contact.

2 3.57 Ω is the resistance of this bulb. Any low resistance indicates a 'good' bulb. If the filament is broken the scale will blink (on most multimeters) indicating infinite resistance, i.e. a break.

Checking DC voltage
(drawing G)

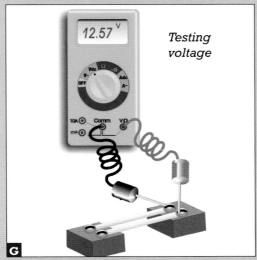

Testing voltage

1. Switch the circuit 'ON'.
2. Set the multimeter to Volts DC.
3. Put the probes onto the two points at which you wish to measure the voltage (red to positive & black to negative).
4. A minus sign in front of the voltage indicates that the red probe has been put on the negative terminal.
5. Read the voltage.

Checking AC voltage
- **Do not expose live AC circuits unless you really know what you are doing. An error can be fatal.**
1. Switch the circuit 'ON'.
2. Set to Volts AC (V~).
3. Hold the RED probe to the live wire (terminal) and the BLACK probe to the neutral wire. The meter will read the volts at that point.

Checking current
(drawing H)
- Boat owners will not normally need to measure current in a circuit, but a 'clip-on' meter is useful to check that an alternator is generating power.
- If a multimeter is used to measure current, the circuit will have to be broken so that all the current flows

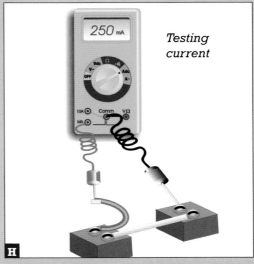

Testing current

through the meter. It will measure only direct current.
- A multimeter will measure only small currents, as the current has to pass through the meter and the probe wires so if in doubt, don't try and measure current.

1. Select Amps DC.
2. Put red wire into mA socket or 10 A socket on the meter according to current expected. If in doubt, start with 10 A.
3. Switch circuit 'ON'.
4. Put one probe on the wire, the other on the terminal.
5. The act of measuring the current will alter the current flow, so it's only an approximation.
6. The non-electronics user has little need to measure current.
7. Measuring alternator output needs a dedicated ammeter. A clip-on meter, although not very accurate, is ideal for trouble-shooting (photo I).

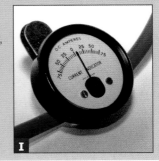

32 Energy consumption

Power
- The power that a device uses is stated in watts (W).
- 1000 watts is 1 kilowatt (kW).
- The current flowing through the device multiplied by the voltage of the supply gives the power in watts.

So:
Power (watts) = amps x volts

AC Shore Power
- The normal shore supply in the UK is limited to 15 amps, though 30 amp supplies are also available, using a different design of plug and socket.
- At 240 volts, this limits the total AC power consumption to 3600 watts (3.6 kW).
- Using multiple devices, i.e. fan heater, water heater, kettle, toaster, etc., may trip the boat's breaker and may even cause the pontoon circuit breaker to trip, cutting off all the other boats as well.
- Read the labels on the appliances to find what you can use at the same time.
- Its probable that the fan heater and the kettle used together will trip the boat's breaker.

DC Boat Power
- Electronic equipment normally has its electrical requirement given in milliamps or amps.
- Lights are normally rated in watts, so to find the current flow, you need to divide the watts by the volts.
- If the power is coming from the battery, the power consumption we are interested in is how many amps for how many hours.
- Multiply the current flow by the time that it is in use and we have its power consumption in amp hours (Ah).
- Add up all the amp hours for all the bits of equipment and we have the total power consumption.

Battery capacity
- Battery capacity is given in amp hours (see the label on the battery). This is a measure of the energy stored in the battery.
- If we know the total energy consumption per day (amp hours per day), we can find out how many days we can run them before we need to recharge the battery.

Estimating daily energy consumption
- Everything you 'switch on' consumes electrical energy, even the panel lights and voltmeter.
- Add up the amp hours per day of everything you use to see what battery capacity and charging arrangements are necessary.
- Although the fridge is 'ON' all day, the compressor runs only as demanded by the outside temperature and the thermostat setting.
- All other consumption figures are a guess at how long each item will be used per day.

Battery capacity and charging requirements
- You are now in a position to decide what battery capacity you need and how you are going to charge the batteries. The example in the table shows 220 Ah consumed in 24 hours.

d.c. power (12 volt)

	watts	amps	hours	Ah/day
Cabin Light	10	0.8	3	2.5
Nav Lights (3 x 10 W)	30	2.4	8	20
Tricolour (25 W)	25	2	8	16
Instruments (day)	3.6	0.3	8	2.4
Instruments (night)	10	0.8	8	6.4
Radar	25 - 50	2.5 - 4.0	3	7.5 - 12
Chart Plotter (mono)	3.6	0.3	8	2.4
Chart Plotter (colour)	10	0.8	8	6.4
Fridge	50	4	5 - 12	20 - 50
Radio	12	1	5	5
CD player	12	1	5	5
TV (12" colour)	50	4	3	12
VHF (receive)	3.6	0.3	8	2.4
VHF (transmit)	60	5	0.1	negligable
Autopilot	25 - 72	2 - 6	5	10 - 24
Heater	60	5	4	20

a.c. power via inverter

	watts	amps	hours	Ah/day
Laptop	50	4	3	12
Microwave oven	600	50	0.4	20
Total per day				220

If you used 220 Ah per day you would need to run your engine (normal regulator), or be on shore power, for at least 12 hours to recharge your batteries.

a.c. power (240 v)

	watts	amps
Microwave oven	600	2.5
Kettle	2000	8
Toaster	700	3
Water heater	1500	6
Battery Charger	360	1.5
Fan heater	2000	8

Note: Shore power supply is often only 13 - 15 amps total. A kettle and fan heater on together can trip the supply.

33 Batteries

- The amount of energy stored in a battery is known as its capacity. It is measured in amp hours (Ah). Stated very simplistically a 100 Ah battery can give 100 amps for 1 hour or 1 amp for 100 hours – or anything in between. The 'amp hour rate' is the current the battery produced when the manufacturer measured the capacity at a specified current. In reality the capacity will vary according to how much current it's supplying, how many times it has been deep cycled, how old it is and other factors. Its capacity will reduce with age.

- The state of charge of a battery is the amount of electricity the battery holds at any given time, expressed as a percentage of its capacity, and is very difficult to measure accurately.
- Batteries can be of 'wet cell', 'gel', and 'reinforced plate' construction. They may be 'vented' or 'sealed'. Many 'maintenance free' batteries do in fact need to be topped up. Only sealed batteries are maintenance free.
- Batteries should never be discharged below about 50% state of charge, otherwise their lives will be severely reduced.

BATTERY MAINTENANCE

Difficulty Rating:

Caution
- Batteries give off hydrogen gas, which is explosive when mixed with air.
- Battery compartments must be ventilated to allow the buoyant hydrogen to escape to the atmosphere.
- Protect against sparks or flames in the battery compartment.

B

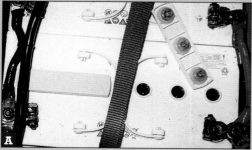

A

1 Any battery that has detachable covers over the cells needs to be 'topped up' with distilled or de-mineralised water (photo A).

2 Inspect monthly and top up if necessary during the summer, or in periods of heavy discharge/charge (photo B).

3 Clean battery cases with a sodium bicarbonate solution, to neutralise acid.

4 Clean battery connections, grease with petroleum jelly (Vaseline), and keep the connections tight.

5 When removing battery connections, prise the jaws apart a little before you pull them off. Pulling a tight terminal off the battery post can pull the post out of the battery.

6 Keep batteries secure.

7 Charge batteries regularly during periods of lay up. (Batteries will self-discharge when not in use. Discharged batteries will quickly become useless and a part-charged battery is susceptible to freezing.)

TYPES OF BATTERY

Engine-start batteries

These must give high current but only for a short period of time. They have thin 'plates' to provide a high current. They are not tolerant of being discharged for long periods, even at lower currents. Starting batteries have a high 'cold cranking amps', at least 450 amps, but do not need high 'capacity'. A single battery of say 60–70 amp hours is normally fine provided its 'cold cranking amps' is high enough, but check the engine handbook. This is the type of battery you will find in your car.

Service or domestic batteries

These are required to give a much lower current, but over a long period of time. They have thicker 'plates' and so can't deliver high peak currents as required for engine starting. Capacity is all important for service batteries as is their suitability for repeated 'charge – discharge' cycles.

General purpose batteries

Some batteries are suitable for both engine starting and service applications and can be a good choice for a cruising yacht. However, like all compromises, they are excellent for neither purpose.

Deep cycle batteries

True deep cycle batteries have a more robust construction. Their advantage over a normal battery is that they allow discharge to 50% for a greater number of cycles. Typically these batteries will stand several hundred cycles (rather than several tens of cycles) before the battery capacity is severely reduced and can no longer 'hold its charge'. They will not tolerate discharging to 'nearly flat'. They are not suitable for starting the engine.

Traction batteries

These very heavy-duty batteries are the only type that will tolerate true 'deep discharge'. They are designed for golf carts and the like, are very expensive and can't be used for engine starting.

Record the date the battery was last topped up (photo C).

STATE OF CHARGE OF BATTERY

This is the amount of energy the battery holds at any given time, expressed as a percentage of its fully charged state, and is very difficult to determine accurately.

A voltmeter will never give the state of charge unless the battery has been 'at rest' for many hours. 'At rest' means that the battery is neither being charged nor discharged, i.e. it is isolated from the circuits.

Battery 'fuel gauges', a voltmeter with a different scale, need the battery to be at rest for at least 3 hours to have any meaning at all. A voltmeter, read in conjunction with an ammeter so that the battery's state of charge can be estimated from the battery voltage and the current flowing at that time, is a much better tool.

A table as shown below can also give a good indication of state of charge and needs to be used in conjunction with a voltmeter and an ammeter (diag. D).

A sophisticated battery state meter will measure the quantity of electricity leaving and entering the battery and so give a reasonably good indication of the state of

E

charge. It will learn the characteristics of your battery, but is expensive (photo E).

Except for very heavy and expensive 'traction batteries', batteries should not be allowed to discharge below 50% as their life will be reduced drastically. This applies to 'deep cycle' batteries as well – they can just be taken down to 50% a greater number of times (diag. F).

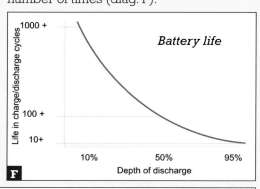

F

Battery 100% state of charge

BATTERY STATE OF CHARGE	BATTERY VOLTS			
	RESTED	0 AMPS	5 AMPS	10 AMPS
100%	12.8	12.5	12.4	12.2
90%	12.7	12.4	12.3	12.1
80%	12.6	12.3	12.2	12.0
70%	12.5	12.2	12.1	11.9
60%	12.4	12.1	12.0	11.8
50%	12.3	12.0	11.9	11.7
40%	12.2	11.9	11.8	11.6
30%	12.1	11.8	11.7	11.5
20%	12.0	11.7	11.6	
10%	11.9	11.6		
FLAT	11.8	11.5		

D

BATTERY CHARGING
- Ideally, all batteries being charged by one source should be of the same type.
- A standard alternator regulator will never recharge a battery 50% discharged to above 75% in any reasonable time, because the charging current falls rapidly to prevent overcharging the battery. A typical 55 amp alternator will start charging at around 40 amps but will be charging at only 10 amps after an hour. It will put into a half-charged battery only about 20 amp hours in the first hour of charging. Thus in a couple of hours of engine running, a 50% charged battery will at best have been charged to 75% full charge.
- An alternator will deliver a very low output with the engine at idle. To deliver a meaningful charge the engine needs to be run at around 50% maximum rpm - see the alternator output curve (diag. G).
- The first hour of charging is the most productive. Each successive hour delivers a considerably reduced charge as the charge current falls, despite the rising voltage – see the fall-off in output curve (diag. H).
- Boats that spend most nights

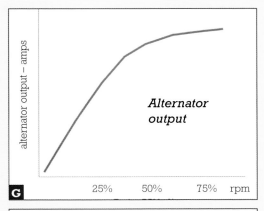

Alternator output

25%   50%   75%   rpm

G

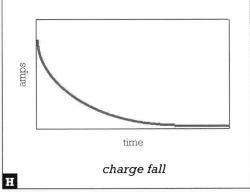

time

charge fall

H

connected to mains power in a marina are served well enough by a standard regulator.
- Boats with heavier electrical demands can benefit from a more sophisticated charging system.

ISOLATING BATTERIES
- The engine start battery must be capable of isolation from the rest of the system so that it's always fully charged when required for engine starting.
- The simplest method of combining two or more 'banks' of batteries for charging is by manual switching.
- Do not select 'OFF' when the engine is running or alternator damage is probable.
- Selecting '1' or '2' powers all circuits from the selected battery bank.
- Selecting 'BOTH' powers all circuits from both batteries.

- Always use the same battery for engine starting. Because it's always fully charged and not cycled it should not suffer from premature failure. However, this will probably result in earlier failure of the 'domestic' battery due to its increased cycling - but you will be able to start the engine!
- Others suggest alternating battery banks for starting and domestics. However, I believe in a dedicated engine start battery. A battery always used only for starting will last 10 years plus and should be much more reliable for that job and it will always be fully charged.

SMART CHARGING

Adding more batteries is not the answer to inadequate battery capacity, unless they can be charged properly. A rough rule of thumb is that the service battery capacity should not exceed 3 times the alternator output, e.g. a 55 amp alternator can charge 165 amp hours of battery capacity (don't count the engine start battery as this should remain almost fully charged at all times). However, charging 200 amp hours with a 55 amp alternator will normally be OK.

A 'smart' charger (photo I) allows a standard alternator to charge a battery much more fully and much more quickly because it prevents the battery being overcharged. A typical 55 amp alternator could then bring a 50% charged 100 amp hour battery to 80% charged in not much more than an hour.

Charge current is then maintained until the battery is 75% charged and then the current falls until the battery is around 90% charged. After this the charge is equivalent to a 'trickle' charge (diagram J).

A 200 amp hour battery bank could be bought from 50% to 80% in about 2 hours, provided the engine rpm was high enough.

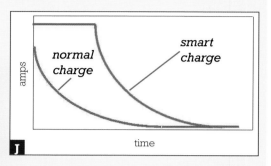

Although fitting a smart charger is not difficult, some alternators require connections to be made inside. This can lead to expensive mistakes and professional fitting may save you money.

MANUAL SWITCHING

- DO NOT SWITCH 'OFF' WITH THE ENGINE RUNNING – alternator failure is likely.
- A rotary switch is commonly used (photo K).

- Separate battery switches are an alternative option (photo L).
- Select 'Engine start' battery for starting,

because coupling to a heavily discharged domestic battery will pull down the engine start battery voltage. It may then be impossible to start the engine.

- Delay selecting 'Both' until the engine battery is charged. This will need only ten minutes or so if the engine battery is kept fully charged.
- Select 'Both' to continue charging the complete system. If motoring for more than an hour or so, select 'domestic' battery, to prevent overcharging the engine battery and ensure as full a charge as possible of the domestic battery.
- As soon as you stop the engine make sure that 'Domestic' is selected.

AUTOMATIC SWITCHING

- Automatic combining of batteries for charging using a blocking diode (photo M) or split charge relay removes the human element that may allow the engine start battery to be inadvertently discharged.
- Charging losses do occur.
- When installing a blocking diode the alternator's voltage sense lead must be attached to the domestic battery terminal. This may entail alteration to the alternator's internal wiring.
- Automatic switching is more expensive.

CHARGING BY SHORE POWER

- Battery chargers should be of the 'smart' variety, offering three-stage charging if a depleted battery is going to be fully charged during an overnight marina visit.
- They must be capable of supplying the evening's needs as well as charging the battery.
- Size will be determined by the domestic load and the size of the battery bank.
- They should be installed in a dry, well-ventilated position. A damp cockpit locker is a poor place for AC power components.

HOW LONG TO CHARGE THE BATTERY?

- This depends on the size of battery bank, the power of the alternator and whether a 'smart' charge regulator is used or not.
- Using the diagram below we see that to charge a 200 amp hour battery that is 50% charged to 90% charged will take 8 hours with a smart alternator and an adequate engine rpm (50% rpm minimum). It will take only 2 hours to raise the state of charge to 80%.
- The first hour or so of battery charging is the most productive.
- We put more charge in an hour into a battery that is 50% charged than one that is 80% charged.
- When cruising, keeping the battery between 50% and 80% charged (rather than trying to keep it as fully charged as possible) will reduce engine running time, especially if you have 'smart' charging', although it will reduce overall life of the batteries.

- A standard alternator will take 6 hours to charge the same battery from 50% to a bit less than 70% and further engine running will add little charge.
- A shore powered battery charger could raise the state of charge from 50% to over 90% during an overnight stay.

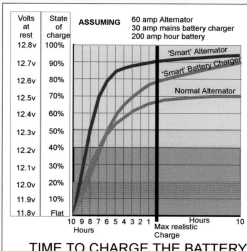

Volts at rest	State of charge	ASSUMING	60 amp Alternator 30 amp mains battery charger 200 amp hour battery
12.8v	100%		
12.7v	90%		
12.6v	80%		
12.5v	70%		
12.4v	60%		
12.3v	50%		
12.2v	40%		
12.1v	30%		
12.0v	20%		
11.9v	10%		
11.8v	Flat		

TIME TO CHARGE THE BATTERY
Time to charge from 50% to 80% = 2 hours using 'Smart' alternator

34 Circuit protection

SHORT CIRCUITS

- Short circuits allow very high currents to flow (a DC circuit could allow 800 amps) and can result in an electrical fire.

- Short circuits result from insulation breaking down, caused by chafing, ultraviolet light or contamination.

FUSES AND CIRCUIT BREAKERS

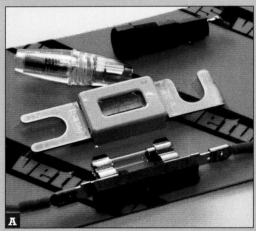

A

- Equipment often has an internal fuse for self-protection. It's usually of a low value.
- Wiring supplying the equipment needs to be protected by a fuse or circuit breaker in case of a short circuit.
- The amperage of the fuse or circuit breaker must never exceed the current rating of the wire.
- All distribution wires must be fused (including the battery) where there's a chance of positive and negative wires coming into contact - equipment cases can be connected to negative.

B1

B2

Screw-in panel fuse.

POSITIONING OF FUSES

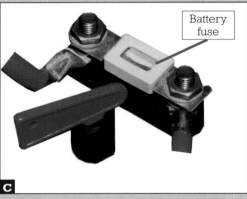

Battery fuse

C

- Ideally, battery fuses should be fitted as close as possible to the battery terminal, but outside the battery box so that in the event of the fuse blowing, any spark will not ignite hydrogen that may be present in the battery box (photo C).

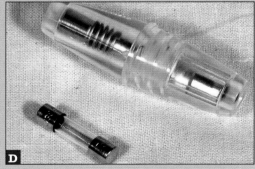

D

Any fuse protecting a critical circuit should be easily accessible, therefore I don't like inline fuses (photo D) that are hidden behind panels, as is commonly the case.

- All wires should be protected by fuses as close as possible to their supply terminals.

E

- Rather than trying to connect lots of new circuits to a single existing terminal, install a new busbar for all additional circuits (photo E).
- Fit panel-mounted fuses or circuit breakers, connected to the busbar (photo F).

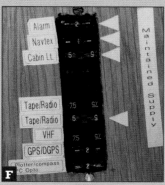

F

G

H

I

- You can also fit a fuse box (photo G) using car 'plug-in' fuses (photo H).
- A car lighter-type socket is often protected by a relatively high value fuse or circuit breaker. If the wire supplying the accessory plugged into the socket is of low current capacity, the plug should be fused, or an inline fuse fitted (photo I).

Installing new equipment

1. Check wire size required, according to its length and the current it's to carry, from the table on page 85.
2. Decide if you will need to power the equipment from the battery (e.g. a VHF radio) or the panel (e.g. a new GPS).
3. Run a new positive wire from the circuit breaker panel or the battery to a new busbar.
4. Run a new positive wire from the busbar to a fuse holder (or circuit breaker).
5. Run a new positive wire from the new fuse holder to the new equipment.
6. Fit an appropriately-sized fuse as recommended by the equipment manufacturer.
7. Run a new negative wire from the existing negative busbar, or a new one as appropriate.

- If you have an ammeter, its shunt should be fitted in the negative battery cable.
- Equipment supplied direct from the battery MUST have the negative cable connected via the shunt and NOT taken direct to the battery's negative terminal.

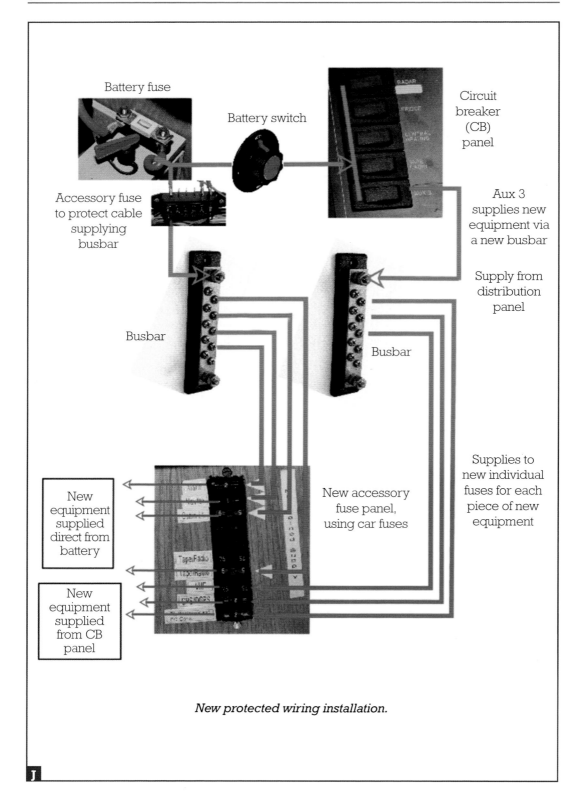

Battery fuse

Battery switch

Circuit
breaker
(CB)
panel

Accessory fuse
to protect cable
supplying
busbar

Aux 3
supplies new
equipment via
a new busbar

Busbar

Supply from
distribution
panel

Busbar

New
equipment
supplied
direct from
battery

New accessory
fuse panel,
using car fuses

Supplies to
new individual
fuses for each
piece of new
equipment

New
equipment
supplied
from CB
panel

New protected wiring installation.

WIRE CURRENT RATING (RATING TABLE)

- The wire must be capable of carrying the maximum current in the circuit. All wires have a current rating. A 5 amp wire must carry no more than 5 amps and should be protected by a circuit breaker or fuse of no greater rating than 5 amps.
- The wire should not allow more than a 3% voltage drop for critical applications and 10% for non-critical ones.
- Normally this is more restrictive than the current rating because it depends on the length (both positive and negative) of the wires.
- Wires bundled together are able to carry less current because they will heat up.

WIRE SIZES REQUIRED FOR A GIVEN LENGTH OF CABLE RUN
(Length is sum of positve & negative wires)

WIRE SIZE 3% voltage drop
Critical applications - bilge pumps, nav lights, electronics, etc.

CURRENT (amps) LENGTH	5 A	10 A	15 A	20 A	25 A	30 A	40 A	50 A	100 A
5m	16	12	10	10	8	8	6	6	2
10m	12	10	8	6	6	4	4	2	1/0
20m	10	6	6	4	2	2	1	1/0	4/0
30m	8	4	4	2	1	1/0	2/0	3/0	
40m	6	4	2	1	1/0	2/0	3/0	4/0	
50m	6	2	1	1/0	2/0	3/0	4/0		

WIRE SIZE 10% voltage drop
Non-critical applications - windlasses, cabin lights, etc.

CURRENT (amps) LENGTH	5 A	10 A	15 A	20 A	25 A	30 A	40 A	50 A	100 A
5m	18	18	16	16	14	14	12	12	6
10m	18	16	14	12	10	10	8	8	4
20m	16	12	10	8	8	8	6	4	2
30m	14	10	8	8	6	6	4	4	1
40m	12	8	8	6	4	4	2	2	2/0
50m	10	8	6	4	4	2	2	1	3/0

American Wire Gauge 'Boat Cable'

AWG	18	16	14	12	10	8	6	4	2	1	1/0	2/0	3/0	4/0
Sq. mm	0.8	1	2	3	5	8	13	19	32	40	50	62	81	103
Max. amps	20	25	35	45	60	80	120	160	210	245	285	330	385	445

Reduce current by 15% when run in engine compartment.
Standard UK wire size - 1, 1.5, 2.5, 4, 6, 25 & 40sq mm.

35 Connections

WHY IT NEEDS TO BE DONE
Poor connections may need to be remade, or you may be adding some more wiring.

WHEN IT SHOULD BE DONE
As required.

TOOLS NEEDED
Wire cutter, wire stripper, crimping tool, heat gun.

Difficulty Rating:

CONNECTIONS
- Corrosion is the enemy of all cable joints on a boat (photo A).

- Although connectors from the local car shop are cheap, they aren't designed for the rigours of a marine environment.
- Proper marine terminals can be obtained from specialist suppliers such as Index Marine, Ancor or Blue Sea. Their products are available from good chandlers or specialists such as Merlin Equipment.

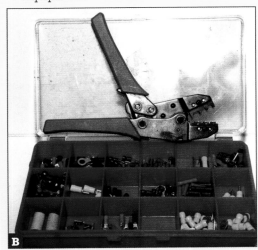

- Always keep a good supply of assorted terminals and use a professional crimping tool (photo B).

Terminal Blocks
- 'Chocolate box' connectors are not ideal for use on a boat.

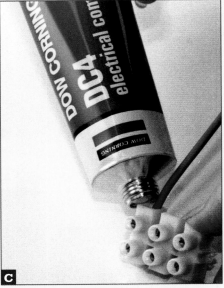

- If used, seal the joints with silicone grease, to prevent corrosion of the cable (photo C).
- Marine-grade busbar connectors should be used with self-sealing crimped terminals (photo D).

Splices

E

- It is best to use heat-shrinkable self-adhesive cable splices (photo E) – OR...

H1

H2

- Alternatively, use normal crimped connectors and apply heat-shrink sleeving (photos H1 & H2) – OR...

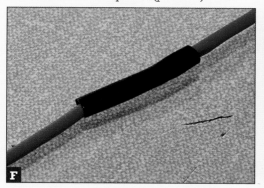

F

- ...soldered joints covered with 'heat shrink' self-adhesive sleeving (photo F).

I

- ...apply silicone grease to the open terminal (photo I).

Crimped Terminals

- The cheap crimping tools supplied in a set with some terminals are not very good at crimping. Much better is a more expensive crimping tool with a clutch – it won't release until enough pressure has been applied.

G

- It is best to use heat-shrinkable self-adhesive cable terminals (photo G).

Instrument (signal) wire joints

- Use 'Eton 23' or 'Scotchlok' type connectors which are gel filled and seal automatically when the joint is made (photo J).
- These can be obtained only from specialist suppliers. However, a local electronics installer may be able to provide them.

J

CRIMPED CONNECTIONS

1. Choose the appropriate connector – the colour is coded for wire size.
2. Don't use a connector too big for the wire – it won't grip properly.

Strip off only enough insulation from the wire so that, when it's inserted into the connector, the remaining insulation abuts the connector (photos K2 & K2).

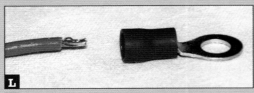

4. Insert bared wire into connector (photo L).
5. Squeeze the connector with the crimping tool to clamp the connector to the wire. Crimp connectors are colour coded according to wire size – use the same colour crimp tool aperture as the connector (photo M).

SIGNAL WIRE CONNECTORS

1. Don't strip the wire.
2. Insert wires into connector.
3. Squeeze the top and the bottom of the connector with a pair of pliers. The connector will pierce the insulation to make a good connection and seal the joint automatically with grease (photos N1, N2 & N3).

HEAT SHRINKING

1. Put heat shrink tubing over the joint.
2. Application of heat using a hot air gun (paint stripper) shrinks the tube to grip the wire and joint tightly (photos O1 & O2).

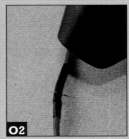

- Hair dryers don't produce enough heat, but a hot-air paint stripper, held not too close will do the job.
- Some heat shrink tubing is coated internally with adhesive that melts when heated. This makes a very corrosion-proof joint.

36 Soldering

WHY IT NEEDS TO BE DONE

Old joints need to be remade, or new joints introduced.

Soldered joints are mechanically strong, but not always appropriate:

- Overheating of a joint (though fire or short circuit) can cause the solder to melt and a live wire become unattached and dangerous.
- Where the soldering ends, there's a stress-raising weak spot on the wire, which may fracture due to vibration. Different people hold different opinions as to whether a wire should be 'tinned' at its end prior to inserting it into a screw-connector.

WHEN IT SHOULD BE DONE

As required.

TOOLS NEEDED

- Rosin flux cored solder (not acid flux, which will cause the wire to corrode and break). (Photo A.)
- A soldering iron not too big for the job, but big enough to bring the components to the correct temperature.
- Several types of soldering iron can be useful on a boat: a small propane iron; a 25 watt 12 volt iron; and a 50/100 watt

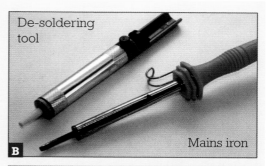

De-soldering tool

Mains iron

12 V iron

Gas iron

mains iron (photos B & C).

- Cheap gas irons don't last long, as their catalytic element quickly fails. It's worth getting a professional one if you intend it to get much use (photo D).
- It can also be fitted with a rope cutting 'hot knife' to seal the ends of cut rope (photo E).

Difficulty Rating: ▪▪▪□□□

(Practise on some old wires first.)

• Allow sufficient time for the iron to heat up fully.

F

1 Clean the tip of an electric iron on a damp sponge or cloth (photo F).

2 Clean a very dirty tip with a file.

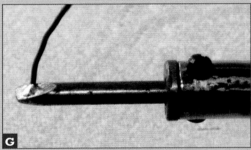

G

3 Apply some multi-core solder to the cleaned tip of the hot iron (photo G). (If you're using a propane torch omit this step.)

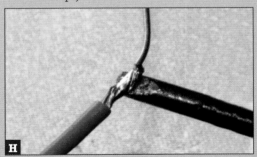

H

4 Heat the wire with the iron then apply some solder to the wire (not the iron) – this is called 'tinning' (photo H).

5 'Tin' the component you are soldering onto.

6 Hold the components together and heat them with the iron. They should join together. Let them cool before you release them.

7 If possible twist wires together before soldering to provide mechanical strength – but you won't then be able to tin the individual wires.

I

8 If you need more solder, apply it to the components, not the iron (photo I).

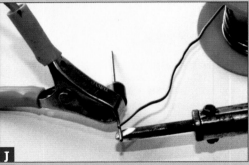

J

9 If you're soldering components that may be damaged by the heat, use a 'heat sink' such as a small 'crocodile clip' to divert heat away from the delicate component (photo J).

 TIP Don't put a hot iron down. Hang it up by its hook or put it in a stand. This avoids burning furniture or the work bench (photo K).

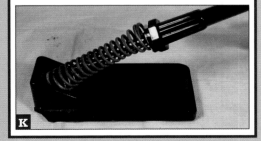

K

37 Wiring

WHY IT NEEDS TO BE DONE

Old wire will, at some time, need to be replaced. Boat owners often add new pieces of equipment and so need to run new wires to supply it with power and to send information to other pieces of equipment.

WHEN IT SHOULD BE DONE

As required

TOOLS NEEDED

See sections on *Connections* (from page 86) and *Soldering* (from page 89).

Electrical wire used on boats should be tinned along its whole length. This is expensive so most production boats have 'automotive' wire, which allows corrosion to spread along the strands of wire under the insulation. This makes it impossible to remake a joint successfully.

Difficulty Rating:

(But you need to keep things tidy.)

A

B

4 Use grommets to protect the insulation where you run wires through bulkheads, etc. (photo C).

1 Do not run the wires through the bilges where their condition will deteriorate (photo A).

2 Use a push wiring threader to run wires through difficult places (photo B).

3 Run a 3mm mouse line too, in case you need to run another cable later.

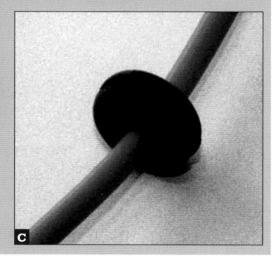

C

5 Don't leave a tangle of wires
   (photo D).
6 Make it all neat, using a new
   distribution board if necessary.
   Then troubleshooting will be easy.
   (photo E).
7 Make a wiring diagram of any new
   work (diag. F).

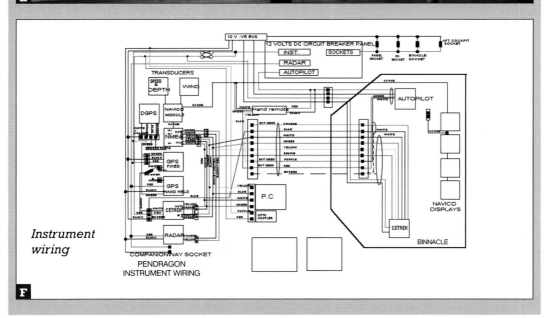

Instrument wiring

PENDRAGON
INSTRUMENT WIRING

38 Troubleshooting

- Troubleshooting an electrical circuit needs a methodical approach.

Alternator not charging – generator warning light 'ON'

1 Stop the engine and check the alternator drive belt.

2 Check the battery voltage.

3 Start the engine.

4 Increase engine revs to about half speed.

5 You should see the voltage rise. The actual voltage will vary according to the battery's state of charge.

6 Unless the battery is heavily discharged the voltage should always be above 13.2 volts.

7 If you have an ammeter that shows the alternator output, a positive reading indicates that the alternator is charging but the actual reading will vary according to the battery's state of charge. A clip-on meter, clamped to the alternator output wire, will show if it's producing power.

- If the battery is being charged, the fault is in the warning light circuit.

- If the battery is not being charged, stop the engine and check all the alternator connections.

Alternator repair is beyond the scope of this book. The fault may also be in the 'smart' regulator, if fitted, although modern versions will revert to the standard regulator if the smart unit fails.

Battery not holding its charge

A battery will self-discharge over a period of time and wet cell batteries should be charged every two months.

A 12 volt battery is made up of 6 cells each giving a nominal 2 volts. If one cell is faulty the other 5 cells will be discharged, trying to hold the faulty cell up. The battery will self-discharge rapidly and will never reach full charge. This is inevitable over time and the more deeply a battery is discharged on a regular basis the sooner this will occur.

If you have more than one battery, it may not be obvious which one is at fault.

1 Charge all batteries fully.

2 Disconnect all but one battery from your system and then use this to try and start the engine.

3 Ensure battery and starter motor terminals are clean and tight.

4 On a sailing boat, turn off the cooling water seacock, to prevent flooding the engine with water.

5 Operate the mechanical stop control to prevent the engine starting. If the engine is stopped electrically (see page 48) apply 12 volts to the stop solenoid.

6 Operate starter for 15 seconds and observe the voltmeter.

7 The battery voltage should remain above 9 volts under the load of the starter motor.

8 Low volts or the starter slowing down indicates a faulty battery.

9 Check each battery in turn.

Faults in an electric circuit

Finding a fault in an electrical circuit entails a methodical approach, but luck can play its part.

- If a light fails to work, the bulb has probably failed, or its contacts have become corroded so check the bulb (photo A, page 94).

- If the bulb is at the masthead, then some continuity checks will be worthwhile first.

- If cable running up the mast have plugs at the base of the mast, these should be well protected and silicone grease applied to the contacts.

- If possible give these joints some form of protection from rain and especially seawater (photo B, page 94).

There are two ways to check circuit continuity

EITHER

Check the wire's resistance between two points.

- Very high resistance indicates a break or poor joint.
- Very low or zero resistance indicates no breaks.

OR

Check the voltage between the positive and negative wires at junctions along the circuit using a multimeter as described in the section on the *Multimeter* (from page 71).

- Voltage at the far end of the circuit should be no less than 90% of the nominal voltage.
- Zero volts indicates a break in a wire.
- A large voltage drop indicates poor contacts or wire that is too small in diameter for the current it is carrying.

Wires that are run up the mast usually have a plug and socket joint at the deck, or have a connection inside the cabin, just below the mast. Deck joints are always vulnerable to corrosion but interior joints can also be affected. These joints should always be the first port of call when something up the mast is not working.

- If there's no voltage at the deck joint, the fault is between the switch panel and here.
- If there is normal voltage then the problem is in the plug or up the mast.

Electrical equipment not functioning

See section on the *Multimeter* (from page 71)

for details of checking continuity and voltage.

Flow process

If several pieces of equipment are protected by one circuit breaker or fuse:

- Check others on same circuit.
- Check circuit breaker not tripped, or fuse not blown.
- Check supply to circuit breaker/fuse – if there's an associated indicator light, its illumination confirms a supply.
- Fuse contacts can become corroded, check that they are clean.

If there's shared wiring for part of the circuit, check other items on the shared circuit.

If only one item is affected and it's a light:

- Check the bulb with a multimeter.
- Check the voltage at the bulb holder.
- Check for corrosion of the contacts.

If it's electronic, and remains dead:

- Check that the power supply is plugged in properly.
- Check the voltage supply at the plug.
- Check the internal fuse.

If there are connections along the circuit's wiring, especially in damp areas or at deck plugs:

- Check these connections carefully.
- Check for corrosion.
- Check the supply voltage.

If the equipment is difficult to reach, such as a masthead navigation light:

- Check continuity of the circuit where it's easiest first.
- Check the supply voltage at the deck plug or where the circuit is broken to allow the mast to be un-stepped.

39 AC power

AC power can kill.
If you don't have experience, use a professional to do work on an AC circuit.

A

- The 17th edition of the UK's wiring regulations was introduced on 1st January 2008. It is illegal for anyone without a certificate of competence to carry out work on AC circuits in homes in the UK, with the exception of very minor work such as replacing a socket. Even this is not allowed in kitchens and bathrooms because of the presence of water. At the time of going to press, it isn't clear how such regulations will affect work on UK registered boats, but the message is clear – tinkering with AC can be dangerous. For boats used on the UK's inland waterways a Boat Safety Scheme Examination will be required, where all electrical wiring will be inspected to ensure that it complies with the regulations.
- Part of the new regulations saw an introduction of new wiring colours in the UK.
- Wiring colours in the USA are different.
- On-board AC should be supplied by a properly-wired on-board system incorporating an earth leakage circuit breaker (ELCB) or residual current circuit breaker (RCCB), NOT an extension lead run from ashore.
- Always connect the shore-side plug last and disconnect it first, in case you drop the 'boat end' into the water.
- Check the polarity (whether the positive is connected to the positive) of the AC connection every time you connect to a new dockside power source. Many dockside supplies have the positive and negative wires reversed. This can be dangerous and can cause corrosion.
- If you have no panel-mounted polarity checker, use a plug-in one in a 13 amp socket (photo A). This can be obtained from your local electrical supplier. It will also indicate an earth fault of European systems, but not of 240 volts supplied from a shore power supply in the USA. The meanings of the lights are shown below.

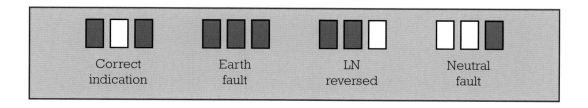

| Correct indication | Earth fault | LN reversed | Neutral fault |

Reverse polarity

- If you find that the marina supply has reversed polarity, inform the marina staff.
- You may find a correctly-wired socket close by.
- Using AC of reverse polarity can be dangerous, as your ON/OFF switches will be in the neutral cable, rather than the live one.
- There may also be implications for galvanic corrosion.

The following will help you use a supply of reversed polarity safely on a temporary basis:

1 Make up a very short cable with a shore connection plug on one end and a socket on the other.

2 DELIBERATELY reverse the polarity of the plug (i.e. connect the brown wire to the neutral terminal and the blue wire to the live terminal on the plug only – wire the socket normally).

3 This will allow you to use an incorrect polarity supply safely, without having to rewire your cable.

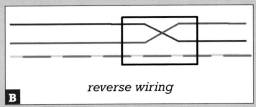

reverse wiring

B

4 Mark this connector cable as a reverser (diag. B).

In countries where 110 volts is the norm, such as the USA, sometimes 230 volts is also available. This is obtained by using two phases of the 3 phase 110 volt supply, to produce 230 volts. In this case, the polarity checker will give a false indication and will have to be ignored.

C

Maintenance

1 Check the ELCB/RCCB trip weekly (photo C).

Do this by pressing the test button – the circuit breaker should 'trip' off, cutting off the power supply.

2 If your breaker keeps on tripping, call in a professional to find out why.

3 Regularly spray the boat's mains power inlet and the connecting cable plug and socket with electrical contact cleaner or water dispersing fluid.

4 If the AC distribution panel is in a damp locker, as in this one mounted in a wet locker on the transom platform, move it. This particular installation caused the owner endless problems (photo D).

D

HEADS

40 Clearing blockages

WHY IT NEEDS TO BE DONE
Blockages can occur because objects other than human waste and soft toilet paper have been put into the toilet bowl. Nautical tradition requires the person responsible to clear the blockage.

WHEN IT SHOULD BE DONE
When it's become blocked.

TOOLS NEEDED
Usually no more than a screwdriver and a pair of rubber gloves.

WHERE TO FIND IT
In the toilet pump.

Difficulty Rating: ☐☐☐☐

First make sure that the seacocks are open and that you are not trying to pump against a closed seacock.

WHALE Mk 5 / HENDERSON Mk IV DIAPHRAGM PUMP

1 Pump through then close the toilet seacocks.

2 Contaminated toilet waste will discharge from the front of the pump.

A

3 Unscrew the pump access cover (photo A).

B

4 Remove the cover (photo B).

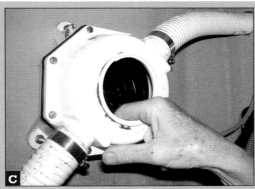

C

5 Check the inlet valve for blockage (photo C).

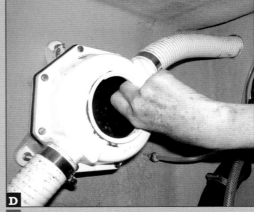

D

6 Check the outlet valve for blockage (photo D).

7 If you need to go further than this, see more on page 103.

ITT JABSCO TOILET PUMP

1 Make sure that the dry bowl
changeover control is fully one side or
the other. If it isn't it will give the
symptoms of a blocked toilet.

2 Close the toilet seacocks.

5 Lift off the pump assembly (photo H).

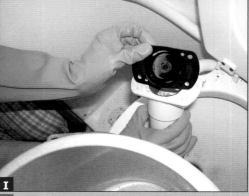

6 Lift off the bottom valve/gasket
(photo H).

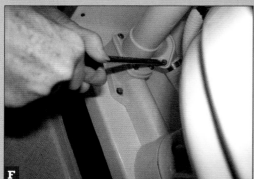

3 Remove the two screws securing the
discharge flange (photos E & F).

Joker valve

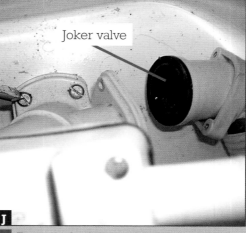

7 Remove the joker valve (photo J).
8 Clear the blockage and reassemble.

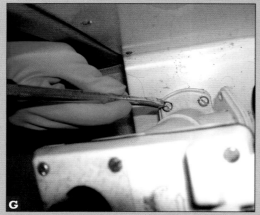

3 Remove the four screws securing the
pump to the base (photo G).

41 Servicing a manual toilet pump

WHY IT NEEDS TO BE DONE

The seal at the top of the toilet pump will wear and eventually leak. The pump itself will get stiff and the valves will become encrusted with lime over time, so a full overhaul will become necessary.

WHEN IT SHOULD BE DONE

As required.

TOOLS NEEDED

Screwdriver and spanners.

WHERE TO FIND IT

Attached to the toilet.

Difficulty Rating: ▮▮▮☐☐☐

Jabsco toilets are the most common (photo A). A leak from the shaft seal at the top of the pump shaft (photo B) can be cured by replacing the seal assembly itself – a spares kit is available.

A full overhaul needs a full service kit (photo C). The ITT Jabsco kit is shown here, others are similar.

Note: The gland design changed in October 1997, so that the pump piston could be withdrawn without first removing the valve cover. Ensure that you have the correct service kit. Pre October 1997 – photo D. Post October 1997 – photo E.

Changing the shaft seal

Once you've removed the assemblies from the toilet, do the work over a bucket to catch the contaminated water and any parts that may drop off.

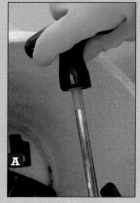

1 Raise the handle to the top of its stroke.

2 Wrap one turn of tape around the piston rod immediately below the handle (photo A).

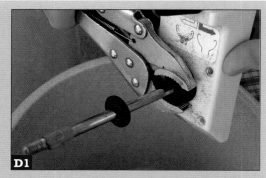

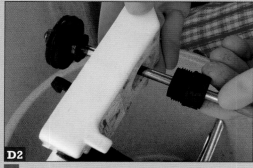

6 Unscrew the seal assembly (photo D1) and draw it upwards off the shaft (photo D2).

3 Grip the shaft with a pair of pliers on the tape (photo B). If you damage the shaft rapid wear of the new seal will occur.

7 Wrap tape around the threaded portion of the shaft to prevent damage to the new seal (photo E).

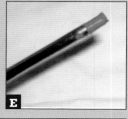

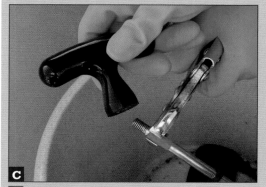

4 Unscrew the handle and remove the rubber bump washer (photo C).

5 Do not let go of the shaft or it may drop into the pump body, you'll then have to take off the valve cover to retrieve it.

8 Fit the new bump washer and carefully slide the new seal assembly onto the shaft (photo F).

9 Remove the tape from the screw thread.

10 Replace the handle on the shaft.

11 Screw the new seal assembly into the valve cover.

Full pump overhaul

1 Flush the toilet so that only seawater remains in the pipework. Pump out the contents of the toilet bowl.

2 CLOSE BOTH TOILET SEACOCKS – if the boat is afloat and the toilet is left disassembled, lock the seacocks closed.

3 When the various joints are opened, any seawater remaining in the system is likely to leak, so be prepared.

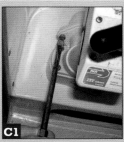
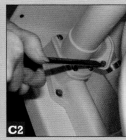

A

4 Loosen the hose clip on the inlet hose (photo A).

5 Remove the inlet hose. If you find this difficult, work can probably proceed with this pipe still attached, provided there's enough slack.

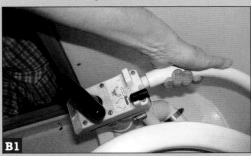

B1

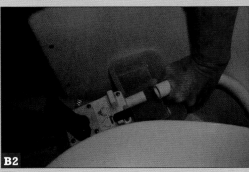

B2

6 Remove the link hose – there's no hose clip and removal should present no problem (photos B1 & B2).

C1 C2

7 Remove the two screws securing the discharge flange (photos C1 & C2).

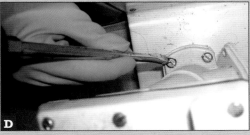

D

8 Remove the four screws securing the pump to the base (photo D).

9 Lift off the pump assembly (photo E).

10 Lift off the bottom valve/gasket (photo F).

E

F

11 Remove the joker valve (photo G).

Joker valve

G

➤

H

12 Slack in the inlet pipe allows the above operations to proceed (photo H).

13 Remove the six screws on the valve cover (photo I).

I

14 Open the 'Flush Control' (photo J).

J

K1

K2

15 Lift the valve cover off the pump body (photos K1 & K2).

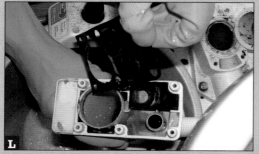

L

16 Remove the top valve gasket (photo L).

M

17 Remove the valve seat (photo M).

N

18 Remove the handle and seal assembly (photo N), as in *Changing the shaft seal*, p100.

19 Remove the piston 'O' ring (photo O).

20 Clean and disinfect all parts.

O

21 Remove scale from cylinder bore using a wooden implement such as a 'lolly stick'.

22 Inspect all rubber components and replace if they are damaged, stiff or covered in scale.

P1

23 Automatically replace the pump shaft seal assembly (photo P1) and the piston 'O' ring (photo P2).

P2

Reverse the procedure to reassemble.

BILGE AND WATER PUMPS

42 Whale Mk 5 / Henderson Mk IV Diaphragm Pump

This pump will be found as the pump on a Blakes Lavac marine toilet and also as a hand bilge pump. When used as a bilge pump, a 'strum box' must be installed at the pick-up end of the suction pipe to prevent debris entering the pipe.

STRIPPING THE PUMP

WHY IT NEEDS TO BE DONE
After a time, the valves will become perished or encrusted with limescale.

WHEN IT SHOULD BE DONE
When the pump becomes difficult to pump or becomes inefficient.

TOOLS NEEDED
Screwdriver.

WHERE TO FIND IT
In the discharge pipe from the toilet and behind the socket for the pump handle.

Difficulty Rating: ▮▯▯▯▯

A service kit is available (photo A).

If it's a toilet pump:
1. Pump through several times to evacuate waste.
2. Close the toilet seacocks.
3. Contaminated toilet waste will discharge from the front of the pump.

Then, for toilet or bilge pump:
1. Unscrew the pump access cover (photo B).
2. Remove the cover (photo C).
3. If it's very tight, you may need to use a wooden block to start it turning (photo D). ▶

E1

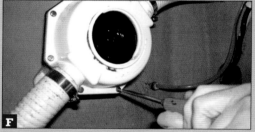

E2

4 Loosen the pipe clips (photos E1 & E2).

F

5 Remove the eight screws holding the pump cover (photo F).

6 Detach the pump cover (photo G).

7 Detach the cover from the pipes (photo H).

G

H

I

8 Hold the centre shaft still, using an 8mm hexagon key, and unscrew the diaphragm retaining nut (photo I).

J

9 Remove the diaphragm and inspect for deterioration (photo J).

K

10 Fit a new diaphragm if necessary and reassemble ensuring that the fittings are the correct way round, smooth sides to the diaphragm (photo K).

11 Remove the inlet valve (photo L).

L

12 Remove the four screws securing the outlet valve (photo M).

M

13 Note that there are two different lengths of screw (photo N).

14 Inspect and replace the valves as necessary.

15 The inlet valve should be smooth so that it seats properly (photo O).

16 The outlet valve must fully seal round the screw holes and the valve. This one is damaged and needs to be replaced (photo P).

17 Check that the outlet valve opens fully, by inserting your finger into the valve (photo Q).

18 If necessary, carefully cut the slots. (I have had to do this on a brand new replacement valve.)

19 Reassemble the valves to the pump cover.

20 Offer up the pump cover and insert all the screws loosely, prior to tightening them all a little at a time (photo R).

21 Inspect the inspection hatch sealing 'O' ring and replace if necessary. Coat the sealing ring with petroleum jelly (photo S).

22 Replace the inspection hatch.

23 Replace the pipes and tighten the pipe clamps.

24 If it's a bilge pump, put some water in the bilge (photo T).

25 If it's a toilet pump, open the seacocks.

26 Check the pump for proper operation.

27 The outer diaphragm is protective only, and its failure may allow water into the boat but will not jeopardise the pump's operation (photo U).

43 Servicing an electric impeller pump

Self-priming water pumps that use a rubber impeller need to be run frequently to prevent the impeller sticking to the pump body. This is especially true for the shower sump drain pump, which may see very infrequent operation. If the circuit breaker trips, or the fuse blows, a sticking impeller is the probable cause.

If freezing conditions are likely, drain the pump body

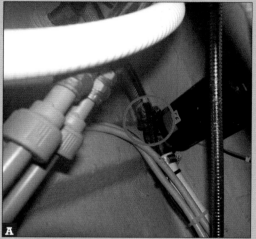

A

1 The pump should be installed so that the drain plug is at the bottom of the pump casing, *unlike* the one shown here (photo A).

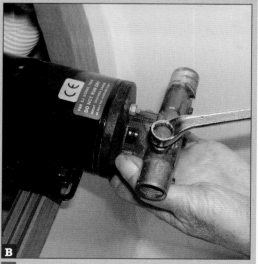

B

2 Remove the drain plug to drain the pump (photo B).

To change or examine the impeller

C

1 Gain access to the pump, removing it if necessary (photo C).

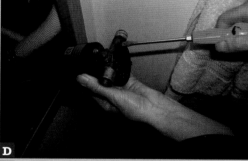

D

2 Remove the faceplate screws (photo D).

E

3 Remove the faceplate, taking care not to damage the gasket or 'O' ring seal (photo E).

4 Note the direction that the blades trail, so that is reassembled correctly.

F

5 Remove the impeller, taking care not to damage the sealing face of the pump (photo F).

6 Switch on the pump to check that it runs, and to make sure that there is not a problem with the pump itself.

G

7 Inspect the impeller for damage and renew if necessary (photo G).

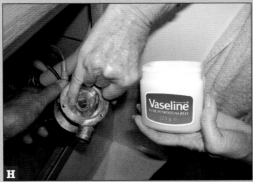

H

8 Grease the pump cavity (photo H).

I

9 Insert the impeller, taking care to have the blades 'trailing' in the correct rotational direction, and grease its front face (photo I).

If in doubt, the direction of rotation of the impeller is 'the long way round' from inlet to outlet.

J

10 Install the gasket, or insert the 'O' ring into its groove (photo J).

11 Replace the faceplate and tighten the screws.

12 Check that the pump operates.

13 Reinstall the pump.

K

14 Put some water in the shower sump and check that the pump works properly (photo K).

SEACOCKS

WHY IT NEEDS TO BE DONE
Seacocks close holes below the waterline. Should the seacock fail, the boat is likely to sink. Additionally, if there's a leak in the pipework and the seacock is seized, it may not be possible to stop the leak.

WHEN IT SHOULD BE DONE
Operate all seacocks regularly, to ensure that they are not seized. Check seacock's condition annually and service the 'Blakes' and 'gate valve' types. Check their associated skin fittings annually for corrosion.

TOOLS NEEDED
Normal tool kit plus grinding paste for 'Blakes'.

WHAT THEY'RE MADE OF
Seacocks for use in salt water should be made from a marine material. Seacocks and skin fittings are sometimes made of brass, an alloy of copper and zinc, and will gradually de-zincify in the presence of seawater, so make sure yours are made of bronze, preferably DZR bronze, which resists dezincification. Seacocks and skin fittings are now available by Marelon in reinforced plastic.

For obvious reasons, skin fittings and seacocks must be serviced with the boat out of the water.

44 Servicing gate valves

A

Gate valves (photo A) have no place on a boat, for several reasons:

- You can't tell visually if they are open or closed.
- Debris can prevent them from closing fully.
- Some of them are not suitable for use in seawater.

However, if you do have gate valves, never leave them screwed fully open as they will be tight to operate. Close them a quarter turn to keep them free, it's also quick to check if it's open or closed because although you will be able to turn the handle either way, it will come up tight in the open direction.

Difficulty Rating:

1 Remove the nut (or screw) retaining the hand wheel (photo B).

B

C

2 Remove the hand wheel (photo C).

D

3 Remove the gland nut on the spindle and ensure the spindle can rotate freely and uncorroded (photo D).

4 Unscrew the main body nut on the stopcock (photo E).

5 Remove the gate assembly (photo F).

E

F

6 Check components for corrosion (photo G).

7 Grease the shaft and thread (photos H & I).

G

H

I

J

K

8 Screw the gate part way up (photos J & K).

L

M

9 Reassemble (photos L, M & N).

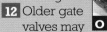

N

10 Tighten packing nut lightly (photo O).

11 If the gland leaks on relaunch, tighten the packing a little more, but don't overtighten.

12 Older gate valves may

O

have replaceable packing in the gland. If the gland leaks and you can't control it by tightening (without the spindle becoming too tight) you will need to repack the gland.

If you have gate valves, consider replacing them with ball valves, or 'Blakes' type seacocks.

45 Servicing ball valves

These usually have a bronze body with a chromium-plated bronze ball set in a Teflon seal. Reinforced plastic seacocks are available.

- There are combined seacock and skin fittings that can be taken apart for annual servicing, although they are less common because of their cost. You may see them in some American boats.
- More common are in-line ball valves screwed directly onto the tail of the skin fitting (photo A). These are not designed to be taken apart and are relatively cheap.
- When the boat is out of the water a stiff ball valve may be greased from outside the hull and, if necessary, the inner pipe removed and the ball greased from this side as well.
- Do not try to clean the ball with a sharp object as the surface can be damaged.
- Regular operation is the best form of maintenance.

TIP On relaunch check all seacocks for leaks.

46 Servicing Blakes seacocks

The traditional Blakes seacock is purpose-designed for marine use (photo A). The latest design has a grease nipple.

Q

Difficulty Rating: ■■■■□□

(Annually, carry out the following service with the boat out of the water.)

R

S

1 Remove the two flange bolts and withdraw the tapered plug (photo R) from the body (photo S).
2 Examine for wear and pitting.
3 Apply cutting paste to the tapered plug (photo T).

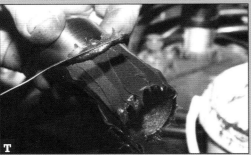

T

4 Insert the plug into the body (photo U).
5 Rotate the handle backwards and forwards while putting a light downwards pressure on the plug (photo V).

U

V

6 Withdraw the plug and check for even grinding marks all round, along the complete contact area, including above and below the side intake hole.

W

7 Apply Blakes seacock grease to tapered plug and reassemble (photo W).
8 Do not tighten the two flange bolts too much as this will expel the grease. Once tightened, if you keep operating the handle back and forth, it will expel more grease and the seacock will become slack. With practice you'll get it just right.
9 When the boat is relaunched, check for leaks.
10 In use, occasionally apply some grease with a grease gun. (This assumes there's a grease nipple, as there will be on newer Blakes seacocks.)

STERN GEAR

47 Cutless bearing

WHY IT NEEDS TO BE DONE

Most boats have a bearing at the outboard end of the propeller shaft. Generally this bearing is made of rubber and will, in time, wear.

Excessive wear will cause shaft vibration and eventually the gearbox suffers damage.

Difficulty Rating:

CUTLESS BEARING

The rear bearing supporting the propshaft is usually seawater-lubricated and made of rubber. It has flutes that allow the passage of water through the bearing (photo A). Make sure the water supply is not obstructed e.g. by a shaft anode mounted too close.

A

If the rear bearing is mounted in the hull 'deadwood' or 'shaft log', water intakes must not be blocked by an accumulation of old antifouling paint (photo B).

B

Some play between the shaft and the bearing is normal but if you can feel movement then it's time to think about replacing the cutless bearing.

'P' BRACKET OR 'A' BRACKET

Where there's a length of propshaft protruding from the hull, the cutless bearing is mounted in a bracket fixed to the underside of the hull. A 'P' bracket has one leg and the stronger 'A' bracket has two. The security of this bracket is of prime importance, because if it comes loose the hull can be punctured.

1 The best time to check the security of this bracket is shortly after the boat has been lifted out and its surface has just dried.

C

2 Try to move the bracket sideways (photo C).

3 Any movement will be shown by water oozing from the joint (photo D).

4 Tighten the bolts on an 'A' bracket.

5 If you see water where the 'P' bracket enters the hull, but you feel no movement, check again next year. If you can feel movement, have a shipwright check it.

water oozing from hull/P bracket joint under sideways pressure
D

6 Refitting a 'P' bracket can be an awkward and very dusty job, as its mounting will have to be ground away inside the boat. Secure refitting is essential and added strength may be required. As such, this may well be beyond the scope of this book.

REMOVAL OF CUTLESS BEARING

- Remove the propeller - see how in section on **Propellers** (page 114).

grab screws

E

1 Cutless bearings are often secured by 'grub' screws (photo E).
2 Scrape paint from the bearing housing to reveal the grub screws.
3 Remove the grub screws, with a hexagonal wrench (Allen Key).

Housing attached to deadwood

deadwood

retaining nut

F

retaining bolt

G

1 Remove the nuts (photo F) or bolts (photo G) retaining the housing.
2 Slide housing off propshaft.
3 Drive out the bearing and replace.

H

Bearing mounted in 'P' or 'A' bracket (photo H)

1 Unless the bearing slides out readily from its housing, with minimal force, this is really a job for a marine engineer.

drift

I

drift moves inwards

J

2 It's not a good idea to drive the bearing out by hammering with a (suitable) drift because the bracket itself could be loosened, unless little force is required (photos I and J).
3 You will need to devise some sort of puller which can be used without withdrawing the propshaft. A marine engineer will have one and you may be able to hire this.
4 If the propshaft has been removed, a threaded puller can be used. An alternative is to saw through the Cutless bearing along its length from inside. You need to be careful not to cut into the bearing housing of the 'P' bracket or it will be seriously weakened.

48 Propellers

WHY IT NEEDS TO BE DONE
Propellers become fouled when immersed, and it's very difficult to prevent this from happening. Propeller fouling causes a very marked drop in performance.

Electrolytic corrosion can cause serious propeller damage

WHEN IT SHOULD BE DONE
At least annually, but any time the boat is

dried out a quick clean of the prop will be worthwhile.

TOOLS NEEDED
Electric drill with a wire brush, abrasive paper and scraper for cleaning the propeller. Spanners, hammer, Allen keys, prop puller or gas blowtorch for removing prop.

Difficulty Rating:

CLEANING PROPELLERS
Apart from fouling, the major enemy of a prop blade is electrolytic corrosion.

1 Thoroughly clean the prop using, in the following order:
2 Paint scraper
3 Wire brush on an electric drill
4 Waterproof abrasive paper

A

5 Examine the blades for 'dezincification'. This appears as a mottled red/brown appearance on the surface, indicating that the zinc in the bronze alloy has been corroded away (photo A).
6 Severe corrosion will show a pitted surface and can weaken the blade significantly.
7 If dezincification is present, check the anodes (see **Anodes**, from page 122).
8 Saildrive legs have relatively small anodes, so dezincification is more likely on heavier, folding props. This could require frequent anode replacement.

Removing variable pitch and feathering propellers
These need to be serviced according to the manufacturer's instructions.

Removing a fixed-bladed propeller
1 Remove the split pin or lock washer and undo the nut a few turns.

B

2 This prevents the prop shooting off if it's difficult to remove and you have to resort to force (photo B).

C

3 If the prop cannot be moved backwards along its taper, you will have to use a prop puller (photo C).

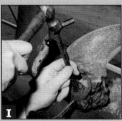

falling out. Remove these (photo H).

2 Drive out each pivot pin (photo I).

4 It's probable that the amateur will not have this tool, so the alternative is to heat the prop so that the hub expands (photo D).

5 Apply heat all round the prop hub (photo E).

6 Gently knock the prop backwards, using a hammer and a block of wood to cushion the blows.

3 Pull the blade out of its socket (photo J).

4 Remove the other blades (photo K).

5 Clean all surfaces thoroughly.

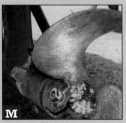

6 Replace the first blade, greasing with waterproof grease (photo L).

7 Align the next blade's gear so that the extension is the same for each blade. Insert the pivot pin and apply grease (photo M).

7 Remove the prop nut and slide the prop off the shaft (photo F). (The prop will be hot.)

8 Make sure that you don't lose the key! (Photo G.)

key

Removing folding propellers

- Grease the folding gear teeth with waterproof grease and check for full and free operation.
- If stiffness is found, remove the blades and clean all surfaces:

1 There are usually locking bolts preventing the pivot pins from

8 Assemble third blade for a three-blade prop (photo N).

9 Check for full, free and synchronised movement of the blades.

10 Refit any locking devices.

11 Fit new hub anode, if applicable (photo O).

49 Sterngland

WHY IT NEEDS TO BE DONE

The sterngland prevents water entering the hull though the propshaft aperture. Its failure can cause the boat to sink. On a lesser scale, as the packing of a traditional sterngland becomes compressed, the gland will leak and seawater will drip into the bilge.

WHEN IT SHOULD BE DONE

A traditional sterngland will need to be adjusted if the leak rate increases. Once it is no longer possible to control the leak, the gland will need to be repacked.

Modern 'dripless' seals normally need no maintenance but may need to be replaced at specified intervals. Most need to be 'squeezed' when the boat is relaunched to expel any air, because seawater is used to cool the bearing.

TOOLS NEEDED

Adjustable spanners or ring spanners according to the type of stern gland, and a tool to remove the old packing. A hobby knife and fine abrasive paper.

Difficulty Rating: ■■■□□

TRADITIONAL STERNGLAND

A

A traditional sterngland with packing, otherwise known as a stuffing box (photo A).

B

cap

C

1. The packing is compressed into a housing, either by a 'yoke' and two bolts (photo B), or by a screwed cap (photo C).
2. Over-tightening will cause overheating and wear on the shaft.
3. Normally you can expect about one to two drips a minute if the adjustment is correct. When the packing is fully compressed it needs to be changed.
4. There's a small clearance between the shaft and the gland housing and water will rapidly enter the boat once the packing is removed so speed is of the essence. It's probably a good idea for the amateur to do this job with the boat out of the water.
5. Undo the retaining bolts of the 'yoke' and pull it forward (photo D) OR...
6. ...wind back the locking nut, unscrew the housing cap and pull it forward (photo E & diag. 1).
7. Remove the old packing. An old spring, opened out and sharpened, makes a good improvised tool (diag. 2).

D

cap

E

8. It's often difficult to remove all traces of the old packing material, but it's essential to do so. A purpose-made 'corkscrew' does a good job if you can improvise one (diag. 3).

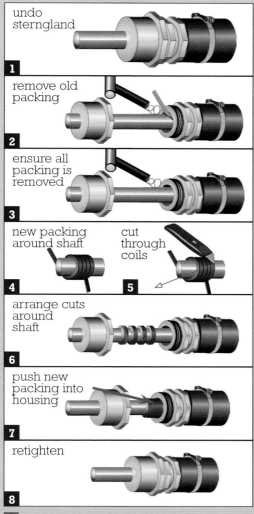

1 undo sterngland

2 remove old packing

3 ensure all packing is removed

4 new packing around shaft   **5** cut through coils

6 arrange cuts around shaft

7 push new packing into housing

8 retighten

9 Clean up the propshaft with some fine abrasive paper.

10 Wind about four turns of the new packing around the shaft (diagram 4).

11 Cut the coils through at 45° with a sharp knife to form four 'rings' (diagram 5).

12 Arrange the rings so that the cuts don't line up (diagram 6).

13 Using some graphite grease, insert the new packing 'ring by ring' into the housing, tamping each one gently home while ensuring that the joins are evenly distributed around the shaft (diagram 7).

14 Slide the compression ring into place and then screw up the adjuster until just tight.

15 With the boat back in the water run the engine in gear for a couple of minutes.

16 You can now check for 'drips' and adjust as necessary (diagram 8).

17 After the engine has been run, the stuffing box should be no more than warm. If it's hot, it's too tight and wear of the shaft will be rapid. Slacken off a little.

Sterngland with greaser

Some sternglands have a greaser, which should be screwed down as required to control the leakage rate.

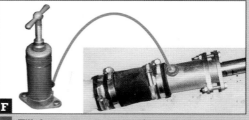

F

1 Fill the grease reservoir as necessary.

2 The packing is often separated by a disc to ensure distribution of the grease (photo F).

Modern 'proprietary' sternglands

These are designed to run without dripping. It's important to observe the manufacturer's instructions on maintenance, especially for sternglands that need to be primed after launching. These generally have rubber bellows and no separate water supply pipe. The Volvo gland, for example, must be changed every 5 years.

HMI's propshaft seal (photo G) has an oil reservoir that needs to be topped up as necessary (photo H).

G

H

STEERING

50 Rudders

WHY IT NEEDS TO BE DONE

Steering failure can be dangerous. Various parts of the system can be subject to severe loads; for example, if someone lets go of the helm with the boat in astern, the mechanism can hit the limiting stops. If the steering is stiff, it will need adjusting.

WHEN IT SHOULD BE DONE

Make a full check of the complete system at least annually. Check the system if an undue strain has been placed on it.

TOOLS NEEDED

None for checking, spanners for adjustment.

Difficulty Rating: ▮ ▯ ▯ ▯ ▯

(For routine checks. Severe damage will require professional rectification.)

Rudder bearings

A great deal of strain is placed on the rudder bearings.

1. Check the bearings for wear by pushing and pulling on the bottom of the rudder to feel for any movement (photo A).

2. Excessive wear means the bearings will have to be replaced and the rudder will need to be dropped from the hull. Seek specialist advice.

Rudder pintles & gudgeons

- Transom-hung rudders normally pivot on pintles and gudgeons (photo B).

- Both the pintles and gudgeons can wear (photo D).
- In this case both sets of pintles and gudgeons were replaced by units constructed by a local engineering shop (photo E).

Rudder skegs

Like the rudder bearings, rudder skegs are under a great deal of stress from steering loads.

1 Remove paint from areas of high stress and check for cracks.

2 Examine the joint between the horizontal and vertical portion of the skeg and the hull itself, to check for any signs of cracking. The cracks may be hardly visible in their early stages.

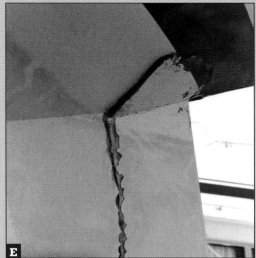

3 Here the failure is obvious (photo E).

4 Skeg failure is normally the result of design or building faults so seek expert advice prior to any repair.

Rudders

1 Scrape back any paint at the top of the rudder to check for cracks (photo F).

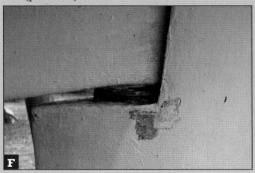

2 Cracks are usually caused by overloading so seek expert advice before attempting a repair.

3 Glass fibre rudders can have very high moisture readings. Water can enter where the rudder stock exits the top of the rudder. If you think your rudder is wet, here's what you can do.

4 Drill a hole (about 5mm) at the lowest point of the underside of the rudder (photo G).

5 This one filled a jam jar with water overnight. It was still dripping a little after a couple of weeks (photo H).

6 Prior to relaunch use a thread cutting tool (a tap) to cut a thread in the drain hole that you have drilled (photo I).

7 Screw a long bolt into the tapped hole (photo J).

8 Put sealing compound under the head and screw tight (photo K).

51 Wheel steering systems

Aft cockpit boats with wheel steering usually have a pushrod system, which is both simple and robust.

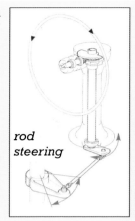

rod steering

- Maintenance involves checking the security of all nuts and bolts (photos A, B & C).

- Where the wheel is attached to the cabin bulkhead, hydraulic steering is common.

- An alternative rotary shaft system is not likely to be found on smaller yachts.

B

A

C

D

1. Regularly check the level of the fluid in the reservoir (mounted just ahead of the wheel) (photo D).

E

2. Top up if necessary with the fluid recommended by the manufacturer (photo E).
3. If frequent topping up is necessary, check for leaks on the steering cylinder. Replacement of the ram seals needs professional attention (photo F).

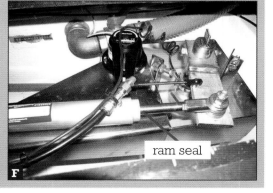

ram seal

F

Where the route of the steering is complex, such as in a centre cockpit boat, the steering system is more likely to use cables.

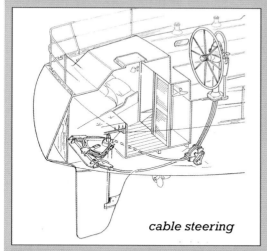

cable steering

I

3 Slacken off the adjusting nuts (photo I).

J

4 Push the cable end though the quadrant as much as possible **by hand** (photo J).

G

1 The cable tension is critical to the feel of the system, and is often too tight (photo G).

K

5 Tighten the adjusting nut as far as possible **by hand** (photo K).
6 Do the same on the opposite side.

H

2 Undo the lock-nuts on the cable adjusters on the steering quadrant (photo H).

L

7 Re-tighten the lock-nuts with a spanner (photo L).
8 Check the feel of the wheel.

ANODES

Sacrificial anodes protect metal immersed in salt or fresh water from electrolytic corrosion.

A sacrificial anode's job is to corrode instead of important metal parts such as the propeller, rudder, drive legs, etc. They are indeed intended to be 'sacrificed'.

Normally corrosion occurs because of the slightly different 'electrical potential' of adjacent, differing, metals. However, in the presence of a poorly-earthed electrical system, either the boat's own DC supply or the mains AC supply, accelerated corrosion can occur. The poorly-earthed mains system does not have to be on your own boat – it may be on another adjacent boat or even the marina system!

WHY IT NEEDS TO BE DONE
Hull anodes will corrode while protecting various external metal fittings.

WHEN IT SHOULD BE DONE
Inspect anodes annually. If you move a boat to a new mooring, especially in a marina where boats are connected to a shore supply of mains electricity, it may be prudent to check the anodes after six months.

If the old anode looks brand new, it isn't working, either because it has nothing to do, or more likely because of poor electrical continuity.

TOOLS NEEDED
Spanners.

Difficulty Rating:					

52 Changing hull anodes

Changing hull anodes
1 The boat needs to be out of the water, although a diver can do the job.
2 Visually check the condition of the hull anodes. They should be able to last until the next check; if in doubt, replace them. They should also not be covered in a dull powdery coating, as this will prevent the anode from working.
3 Make sure anodes are never painted.

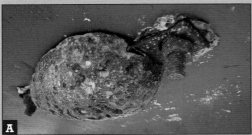

4 Undo the bolts securing the anode (photo A).

5 Remove the old anode.

6 Clean the attachment threads and fit new nuts and sponge pads if necessary (photo B).
7 If the studs are in poor condition they must be replaced (photo C).
8 Fit a new anode (photo D).
9 There must be good electrical contact between the anode and the attachment stud, so wire brush the stud if necessary, and use new spring washers and nuts if required.

step 10 ➤

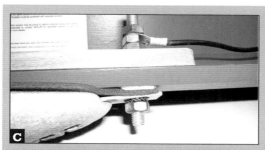

attachment bolt, the engine and the engine end of the propshaft (diag. F).

- Later models of the Volvo Penta saildrive legs are electrically isolated from the engine.

There must be no electrical continuity between any hull anode and the prop or the leg's ring anode.

10 Check the electrical continuity of the anode by using a multimeter. Put one probe on the anode and the other on any exposed unpainted metal such as the propshaft or propeller. The maximum reading should be no more than a couple of ohms (photo E).

- If there's no electrical continuity between the anode and the propshaft, check the continuity between the 'inside the hull'

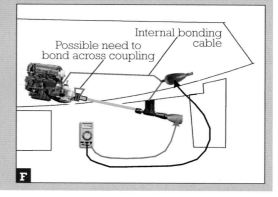

Electrically-isolated gearbox/ propshaft couplings

Some anti-vibration and flexible couplings have no electrical continuity between their driving and driven parts.

If yours is like this you have three options:

1 Fit a shaft anode to protect the propeller. If the propeller is not massive, this may be sufficient. If it is, it will probably have its own propeller anode as well, but often these are woefully inadequate.

2 Fit a bonding strap across the coupling from under a bolt-head on one side to a bolt-head on the other (photo G).

3 Fit electrical brushes in contact with the propshaft (drawing H).

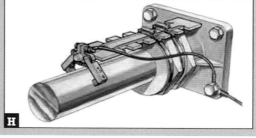

53 Changing saildrive anodes

WHY IT NEEDS TO BE DONE

The underwater drive leg is made of aluminium and corrosion will occur if the protective paint is damaged. The sacrificial anodes on a saildrive leg protect the leg itself and the propeller. These anodes are relatively small in comparison with the mass of the metal they are protecting. Failure to replace these anodes can lead to expensive damage to the leg and the propeller.

WHEN IT SHOULD BE DONE

Corrosion of these anodes can be rapid and they must be checked at least annually. If a large folding propeller is fitted, these anodes may not last a full season.

TOOLS NEEDED

Spanners and screwdrivers of various sizes, mallet and drift.

Difficulty Rating: ■ ■ ■ □ □

The propeller will have to be removed. To avoid losing the propeller when underway, proper engineering techniques must be observed.

Changing saildrive anodes

1 The boat needs to be out of the water.

anode

2 Examine the sacrificial anode for corrosion. It is unlikely that the anode will last a second season and in some cases it will not even last a whole season (photo A).

3 Remove the propeller. Here we illustrate a Volvo Penta 3-blade folding prop, probably the most complex of this type.

- Unscrew the three Allen screws that prevent the hinge pins from dropping out (photo B).
- Drive out the hinge pins (photo C) and remove the blades (photo D).

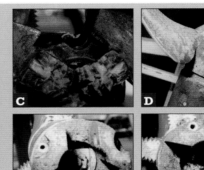

4 Bend back the locking tabs (photos E and F).

5 Unscrew the centre bolt (photos G and H).

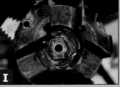

6 Remove the centre retaining nut (photos I and J).

7 Pull off the prop (photo K).

8 Remove the two cross head screws holding the anode in place (photo L).

9 Clean all components thoroughly. Fit the new anode.

10 Replace the rope cutter, if fitted, and grease the propshaft (photo M).

11 Refit the prop hub (photos N and O).

12 Replace the hub nut (photo P).

13 Refit the new tab locking washer (photo Q).

14 Replace the centre bolt and bend up the locking tabs (photos R and S).

15 Grease the blade folding gear teeth and hinge pins, and refit (photo T).

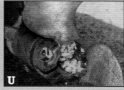

16 Ensure that the gears mesh correctly (photo U).

17 Refit the locking Allen screws (photos V and W).

18 Photo X shows how the hub nut and locking nut fit into place.

19 The pinkish surface corrosion on this propeller hub indicates insufficient anodic protection. This is common on these heavy propellers on saildrive legs, where anodes are inevitably on the small side. The hub is unlikely to be weakened, but eventually the prop blades will be (photo Y).

20 The new ring anode fitted. But it probably won't be there at the end of the season! (Photo Z.)

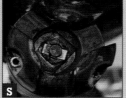

> **TIP**
>
> If the attachment screws are very close to the edge of the anode, the anode may fall off long before it has corroded away. This is because the small amount of anode where it is attached won't last very long. Paint this area of the anode with antifouling paint so that it won't corrode.

54 Changing shaft anodes

WHY IT NEEDS TO BE DONE

An anode should be as close as possible to the metal component that it's protecting. With large propellers which are some distance from the hull anode, extra protection can be given by a shaft anode.

WHEN IT SHOULD BE DONE

At least annually.

TOOLS NEEDED

Screwdriver or small spanner, abrasive paper.

Difficulty Rating: ▮▯▯▯▯

A **B** **C** **D**

1 Remove the old anode (photo A).
2 Thoroughly abrade shaft to ensure good electrical contact (photo B).
3 Fit the new anode about 1 inch (25mm) in front of the cutless bearing to ensure that water will flow into the rubber slots of the bearing (photo C). This is because water lubricates the bearing and if the anode is mounted too close, as often happens, lubrication may be reduced.
4 This one is too close to the 'P' bracket (photo D).

Replacing propeller anodes

Some propellers have anodes attached to the rear of the propeller boss (photo E).
- These are small and often won't last the season.
- If the anode is very thin around the attachment bolts, paint the anode in this area only with antifouling (photo F) to prevent premature wear (or it may fall off).

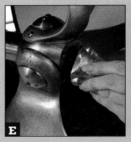

E **F**

TIP Anodes can work loose because the metal next to the shaft corrodes, making the hole in the anode too big for the shaft. The anode will then start to shake about causing noise.

F

Some anodes are cast around a steel armature that stays firmly fitted to the shaft, preventing the anode from shaking loose. It's worth paying the extra for these (photo F).

TIP Stick the small ring anodes to the hub with sealant, but make sure that this doesn't stop the retaining screw coming into electrical contact with the hub. Check with a multi-meter. This helps to stop centrifugal force from throwing off the weakened anode.

ANTIFOULING

WHY IT NEEDS TO BE DONE
Normal antifouling loses its effectiveness over time, either by being eroded or by its ingredients being used up.

WHEN IT SHOULD BE DONE
Most antifouling paint needs to be reapplied annually, although there are a few expensive types that claim to last for two years. Copper antifouling coatings may last for ten or more years but, in all cases, the hull will need to be scrubbed at least annually.

TOOLS NEEDED
Scrapers, paint brushes, rollers and roller trays.

Difficulty Rating: ▮ ☐ ☐ ☐ ☐

REGULATIONS
National regulations governing the active agents in antifouling paints are continually changing under environmental pressures, and manufacturers are being increasingly pressurised into formulating more eco-friendly paints. This also means that this year's paint may not be as effective as last year's. When an antifouling is reformulated, it's required to be renamed. Regulations are also being introduced on disposal of antifouling residue. Boatyards are required to have special run-off pits in the pressure washdown area, and scrubbing down of erodible antifoulings on drying out piles or grids will become a thing of the past. Also, if you are going to scrape off the old antifouling you should collect all the scrapings (on a plastic sheet) for proper disposal.

HEALTH AND SAFETY
Use disposable latex or plastic gloves and cover areas of exposed skin when using and rubbing down antifouling paints. Antifoulings should always be wet-abraded, which rules out electric sanding. Wear safety glasses to prevent splashes entering your eyes. Look for any special warnings on the tins and consult the manufacturer's literature. If there's any chance of antifouling residue flying around, wear a mask or respirator.

TYPES OF ANTIFOULING PAINT
There are three general types of modern antifouling: erodible (self polishing or copolymer); hard; and traditional. Choice depends on the type of sailing you do and where you do it. Local enquiries may give a consensus of which works best in your area.

- Aluminium saildrive legs must NOT be painted with an antifouling containing Copper Oxide, especially if there's any damage to the protective paint system. Only use paint suitable for use on aluminium, such as International's MPX, intended for sterndrives. 'Peller-Clean' from Shogun is a silicon-based antifouling formulated for propellers and outdrives and is claimed to last for at least 3 years.

Steel and aluminium hulls require special antifouling paints and treatments.

Newly-antifouled hulls must be relaunched within a specified time, which can vary from two weeks to more than three months according to the antifouling.

APPLICATION CONDITIONS
Most paints have a minimum application temperature of 5⁰C. If you moisten (with water) an area to be painted and it dries in 15 minutes or less, humidity will be OK for antifouling. Drying / overcoating times will vary with temperature: observe the manufacturer's recommendations. Get their data sheets or painting guides for all the information. Painting onto a damp hull will give poor adhesion.

55 Removal of old paint

Scraping off

1 A scraper of the 'Scarsden' type is effective (photo A), i.e. a handle fitted with a

hardened steel blade. But it is tiring and time consuming and unless skill is used, the corners of the blade can cause damage to the gelcoat.

2 Chamfering the sharp corners can help prevent 'digging in' (photo B).

3 Scrape off the paint (photo C).

Chemical stripping

Several makes of stripper are available, but it's essential that one formulated for use on

fibreglass is used as otherwise serious damage will be caused to the gelcoat (photo D). Applied according to the instructions, stripping is relatively easy and quick, but messy and fairly expensive.

Grinding

Non-encapsulated metal keels can have the old antifouling removed by grinding.

1 Use an angle grinder with a Cintride grinding disk.

2 As there will be dry antifouling dust, the correct protective gear MUST be worn (photo E).

3 Deep rust pockets can be scoured clean with a wire brush attachment on the grinder (photos F).

Slurry blasting

This is the most expensive, pain-free and quickest method of removing old antifouling. A specialist contractor uses high-pressure water mixed with sand to remove the old paint quickly and leave the surface ready for the new antifouling. If you have a cast iron keel, this will be sand blasted to remove all paint and rust and a protective coating applied immediately to prevent corrosion. He will observe environmental regulations as appropriate. Slurry blasting may remove any protective epoxy coating and will damage the gelcoat above the waterline, so ensure that it's well masked. Blasting with solid carbon dioxide pellets is very clean, and won't remove the epoxy or damage the gelcoat, It will leave a finish too smooth for painting so will need to be abraded first.

56 Preparation

PREPARATION
Application directly to gelcoat

1. Use a hull cleaner/degreaser to remove wax, grease or silicone residues.
2. Use 80 to 240 grade dry abrasive paper on a cork block to remove gloss. This should not be required after slurry blasting as the surface will be 'keyed'.
3. Wash thoroughly to remove dust.
4. Allow to dry and check for an even matt finish.
5. Apply a coat of antifouling primer.
6. Allow specified overcoating time before applying antifouling.

Repainting with an antifouling of the same type

Paint manufacturers provide information on the compatibility of different types of antifouling.

The easiest way by far to remove antifouling is by pressure washing (photo G).

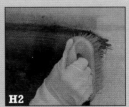

5. Wet abrade the filled areas.
6. Patch-prime any filled area (photo K).
7. Wet abrade the primed areas.
8. Wash off any sanded areas to remove dust.

Application to an antifouling of different or unknown type

1. Proceed in same manner as above.
2. Apply antifouling primer to whole hull.
3. Dry abrade (120 grade) to provide a key for the antifouling paint.
4. Wash down the hull to remove dust.

1. Once the yacht has been chocked off, remove any fouling that has not been removed by pressure washing. The sooner this is done after the boat comes out of the water, the easier this will be. Dried fouling is difficult to remove.
2. Use a scrubbing brush or an abrasive pad with copious amounts of water to remove fouling (photos H1 and H2).
3. Wet abrade (120 grade) any rough areas (photo I).
4. Fill any areas that are deemed to be too rough with a waterproof filler (photo J).

MASKING UP

- Ordinary masking tape should be removed within 24 hours to avoid adhesive sticking to the hull.
- 3 day and 14 day masking tape is available and should be removed as soon as the last coat of paint is touch dry. Special 'flexible' masking tape for use on highly curved areas is useful for the after sections of modern yachts (with a full transom and a bathing platform).
- 50mm tape helps to avoid paint getting onto unpainted areas but is more difficult to use on curved surfaces.

57 Application

- I find application by roller is the quickest and easiest method of application. I find that a 100mm roller, though smaller, is easier to use and no slower overall than a standard roller.
- It's normally recommended that two coats be applied so that you end up with the correct thickness of paint to last the season.
- Put a third coat around the waterline to allow for periodic scrubbing, as the build up of growth is greater here where there's more light (photo A).

- Keep the tin indoors overnight, especially if you'll be working outdoors in cooler temperatures. Warmer paint is easier to apply. Thinning the paint is not normally recommended.

1. Open the tin and stir the contents thoroughly. Settlement of the contents is very likely so it's essential that it's stirred until the consistency is uniform (photo B).

2. Pour some paint into the roller tray (about enough for no more than ten minutes work, especially if temperature is above 15 degrees, as the paint will start to thicken in the tray) and replace the lid on the can.

3. Dip the roller into the paint, then work it over all the roller.

4. Apply paint to hull (photo C), but don't work it out too much as you'll end up with too thin a coat. If you use

significantly more or less paint than the estimate, the paint film thickness is incorrect.

5. Don't apply paint right up to any supports or you will get a thick build up at the edges (photo D).

6. Don't apply paint to any anodes, as this will prevent them from working (photo E).

TOUCHING UP

1. After you've finished roller application, use a brush to apply paint up to any support pads, making sure that the edges have only a thin application to avoid a thick build up (photo F).

2. Put only a couple of thin coats on the depth transducer, especially if there's a lot of copper in the paint.

3. Clean out any openings in seawater intakes (photo G).

4. Arrange with the yard for props to be repositioned after the second coat has been applied (photo H).

5 Apply two coats of antifouling to unpainted areas where the props were (photo I).

6 When the boat is in slings prior to relaunch, you'll need to paint the bottom of the keel. You may not have had the opportunity to clean this up as well as you would have liked, so do the best you can. If you can persuade the yard to have the boat left in the slings over lunch, or better still overnight, it will give you more time.

58 Saildrive legs and keels

- There will be a build-up of antifouling paint on the saildrive leg but rubbing down must be done very carefully so as not to remove the leg's protective paint scheme.
- Make sure that there is no build-up of fouling inside the cooling water intake holes (photo A).

intakes

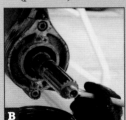

anode

- Apply two coats of antifouling paint suitable for use on aluminium (photo B).
- Do not paint the anode (photo C).

Cast iron keels tend to be porous and usually have embedded slag, which is what may have caused the pitting in the first place. Specialist primers are used for protection of the surface, which must be bright metal prior to application, but unless you are shot blasting this is very difficult to achieve, so application of a rust inhibitor may be advisory. Paint manufacturers won't guarantee their products if you use an alternative treatment such as a rust

inhibitor, but neither do they guarantee that it will stick to rust, so the choice is yours. One such rust inhibitor is Fertan.

1 If you're going to use an anti-rust treatment, now's the time to do it.

2 Follow the instructions on the container (photo D).

3 Spot-prime the rusty patches (photo E), or the whole keel. Apply 5 or 6 coats of steel primer to the whole keel, observing the overcoating times given (photo F).

4 Apply a coat of antifouling primer.

5 Apply two coats of antifouling.

- Brushes and rollers can be wrapped in 'cling film' overnight and will then be ready for re-use without the need for cleaning between coats (photo G). Alternatively wrap them in a plastic bag.
- If you use different colours for each of the two coats, it's very easy to see if you've missed a patch.

FIBREGLASS

59 Repair of puncture damage to the hull

WHY IT NEEDS TO BE DONE
Damage to the fibreglass structure needs repairing to prevent moisture penetrating the lay-up behind the gelcoat.

WHEN IT SHOULD BE DONE
As soon as possible after the damage occurs.

TOOLS NEEDED
Depends on the extent of the damage.

Electric grinder, electric polisher, abrasive paper, cutting compound, polishing compound, brushes, palette knife.

WHERE TO FIND IT
Look for any damage.

Difficulty Rating: ▉▉☐☐☐

(But care and patience are needed to produce an invisible repair.)

If the hull is punctured the damage needs to be tackled from both inside and out. The same applies if crazing is visible.

Where structural integrity is suspect seek professional advice before you tackle the repair.

1 Evaluate the damage and decide if you can tackle the job yourself.

A **B**

2 Gain access to the inside of the hull and temporarily remove any obstacles to the work (photo A).
3 Grind the inner surface to ensure good adhesion of the chopped strand mat (CSM) (photo B).
4 Grind outside to expose the fracture and to remove any crazing (photo C).

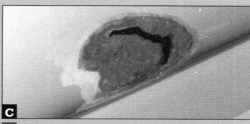

C

5 Ensure protection for your hands, then clean all surfaces with acetone.
6 Mix up resin according to the instructions supplied.
7 Apply resin to the inside surface with a brush or roller.

D1 **D2**

8 Lay up a couple of layers of chopped strand mat (CSM) about 100mm larger all round than the area of damage, with the second layer

D3

larger than the first. Ensure that it is well wetted in and that no edges show a tendency to lift. Don't rush this, take as much time as is necessary. (Photos D1, D2, and D3.)

9 Apply resin to the exterior of the damaged area.

E

10 Apply a couple of layers of CSM to the ground-out areas of damage (photo E).

F

11 Wet well in, but don't bother about the edges. These can be left dry and proud of the surface. The requirement here is to ensure that the fracture is well reinforced so that cracks will not appear after the repair is finished. Ensure that the effective layers of wetted CSM remain inside the final hull contours (photo F).

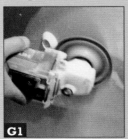
G1

G2

12 After the resin has cured, grind the outside so that the contour remains at least 5mm below the original contour (photos G1 and G2).

H1

H2

13 Mix up some two part marine filler, following the instructions supplied (photos H1 and H2).

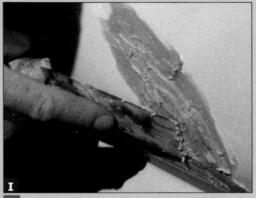

I

14 Fill the damaged area with filler (photo I).

J

15 This time ensure that it stands slightly proud of the original contour. A deep repair may need several layers (photo J).

16 After the filler has hardened, grind it down with a relatively fine abrasive disc so that a smooth surface is obtained and the original contour is preserved. If necessary use a straight edge to ensure the proper contour is achieved (photos K1, K2 and K3).

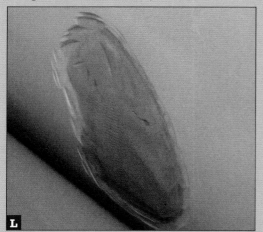

17 Grind down the whole filled area to about 1mm inside the final contour to allow room for the gelcoat. Make a sharp cut at the edge as shown. If you leave bumps they may well allow filler to be exposed at the surface when you rub down the gelcoat and you'll have to start again (photo L).

18 Mix up some gelcoat according to the instructions supplied, stirring well (photos M1 and M2).

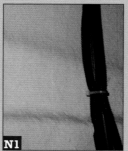

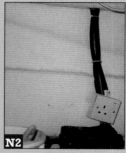

19 If the inside of the hull consists of a gelcoat applied directly to the inside layer of glass cloth, all you'll need to do is to brush matching gelcoat onto your repair without needing to sand down the result. (Photos N1 and N2.)

20 Apply the gelcoat to the external repair. Stipple at first to fill any pin holes, then brush to build up the thickness. About 3 layers will be needed, each one covering a larger area. Allow each layer

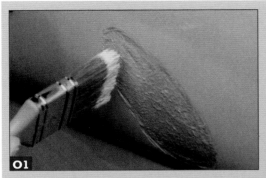

O1

O2

to cure before applying the next.
(Photos O1 and O2.)

P1

P2

21 Rub down the gelcoat with various
grades of abrasive paper, starting with
120 and finishing with 1000. Use a water
spray to lubricate the paper and prevent
clogging. (Photos P1 and P2.)

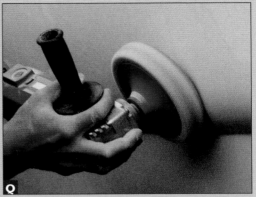

Q

22 Cut down the surface with cutting paste/
liquid, progressing from medium to
finishing grades. Finish off with polish
(photo Q).

R

23 Reassemble the interior components
(photo R).

S

24 Amateurs are unlikely to have all the
tools desirable, but both small and
large angle grinders (photo S) are
useful and can be hired.

60 Repair of stress cracking

WHY IT NEEDS TO BE DONE
Cracks in the gelcoat, looking a bit like a spider's web, are unsightly. More to the point they will allow water to penetrate the laminate, and therefore need to be repaired.

WHEN IT SHOULD BE DONE
As soon as possible after the damage is noticed.

TOOLS NEEDED
Triangular scraper or small hobby electric drill, electric sander, abrasive paper, sanding block, cutting compound and polish.

WHERE TO FIND IT
Keep your eyes open for any gelcoat damage.

Difficulty Rating: ▉ ▉ ▉ □ □ □

Gelcoat colour matching
The most important bit, for any gelcoat repair, is matching the gelcoat colour. Where a repair has been made to the inside of fibreglass moulding, the inside needs to match the interior colour and texture. Ensure this is applied over the whole area exposed so that there's no evidence of the internal repair. This is important to subsequent resale value.

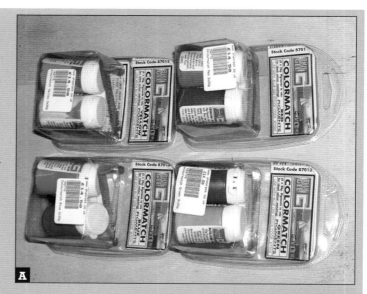

A

External matching may well take several attempts, and patience is essential. The final aim is for an invisible repair. If the boat is fairly new, the builder's standard gelcoat may well give an excellent match. Older boats will have weathered, making matching more difficult, even if you can still get the standard colour. Have a look round and you'll see how many different whites there are and you'll understand the problem. Different colour pigments can be bought to add to the basic colour to provide different shades. For a small job a pack from the chandler will be sufficient (photo A). For bigger repairs consult a resin supplier. With very small jobs on a white hull, a tube of gelcoat filler will probably be nearly invisible. It will in any case change colour as it ages.

Mixing a wax additive with the gelcoat ensures a tack-free (non-sticky) finish where no rubbing down is required. A 'tube' of filler will have this incorporated.

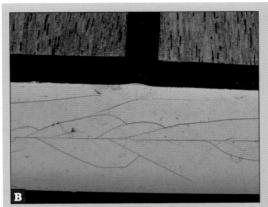

B

Before undertaking any repair of stress cracks (photo B), you first need to find their cause. At a stanchion base, for instance, cracks imply that there is insufficient reinforcing of the deck. A radiating pattern from a point indicates impact damage by a sharp object. Stress cracks in the region of the mast step may indicate settling of the mast support due to insufficient strength of the frames in the region of the keel. Problems like this may be common to all yachts of this model so look at other boats, or make an enquiry to the class association.

If the cause is due to insufficient rein-forcing, increase the strength otherwise the cracking will reappear, (unless the problem is due to accidental collision damage). Expert advice may be needed. Assuming that any necessary strengthening work has been undertaken, the cracks can be filled.

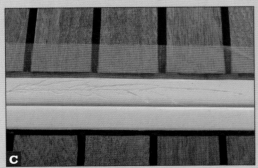

C

1 Protect any exposed woodwork using masking tape (photo C).

2 The cracks must be opened up to allow gelcoat to be applied. Use a sharp triangular scraper to gouge out a triangular trough. Exercise care because the tool will want to wander off the line of the crack.

D

E

3 A better tool is a small electric hand tool such as a 'Dremel' or Black and Decker drill / grinder (photo D) - you have much better control of the tool and it's much quicker. However, its not the sort of thing that everyone has in their toolkit and it probably won't be worth the cost for a small one-off job (photo E).

F

4 Make sure that you have removed all the cracks. You may have to dig quite deep to get to the bottom (photo F).

5 Fill, smooth and finish as for a cosmetic repair.

61 Repair of cosmetic blemishes to the gelcoat

1 Clean the area with a hull cleaner/degreaser to remove any grime, polish or silicone.

2 Rub down the area to be repaired with 120 grade paper (dry) to provide a key for the gelcoat and remove any raised areas caused by the damage (photo G).

3 Measure the correct proportions of the gelcoat filler as instructed with the product (photo H).

4 Mix the gelcoat and hardener in the correct proportions. Don't contaminate either of the tubes with contents from the other (photo I).

5 If it's white, both components will be white, so keep on mixing thoroughly because you won't be able to tell by its colour when it's mixed properly.

6 Apply gelcoat to the damaged area (photo J), aiming to leave the new surface slightly proud (photo K).

7 Allow to set hard.

8 If you need to remove much filler, use a power sander carefully with 60 grade, then 120 grade abrasive (photo L).

9 Rub down the new gelcoat starting with 120 grade and finishing with 1000 grade, used wet (photo M).

12 After a couple of weeks apply boat polish as a protection (photo Q).

10 Keep the surface wet using a spray (photo N).

11 Cut the surface with finishing paste (photo O) and finishing liquid (photo P) to give a gloss surface.

DECK HARDWARE

62 Fitting a deck vent

WHY IT NEEDS TO BE DONE
Ventilation in many boats is insufficient, causing condensation within the cabin.

WHEN IT SHOULD BE DONE
If there's insufficient ventilation.

TOOLS NEEDED
Half inch electric drill, 100mm hole saw (from the local tool hire shop), screwdrivers, hobby knife.

Difficulty Rating: ▮▮☐☐☐

Accurate measurement is the key to successful positioning of the vent, because you need to consider where it should be placed both above and below the deck.

1 Decide where you want the vent outside.
2 Decide where you want the vent inside.
3 Are these compatible? If not reassess where to fit the vent (photo A).

A

4 Consider the likelihood of any internal wiring above the head lining. If you have a removable headlining, remove it to check. If you have a vinyl lining stuck directly to the deckhead, you'll be able to feel any wires. If you have a solid lining bonded to the deckhead, you'll need to check with the builder.
5 Mark where the small (3mm) pilot hole is to be drilled and, with a pair of compasses, draw a circle the size of the cut-out for the vent (photo B).

B

6 Remeasure everything.
7 Drill the pilot hole from outside all the way through to the cabin (photo C).

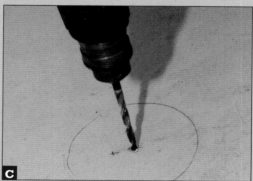
C

8 If the head lining is made of vinyl cloth, mark a circle the size of the vent on a piece of stiff card and cut out the centre a couple of millimetres smaller than the diameter to make a template (photo D).
9 Drill a pilot hole in the centre of the template, locate the hole on the inside of the cabin and offer up the template

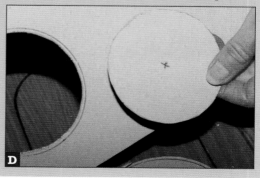
D

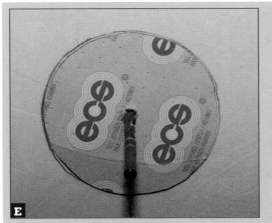

E

using the drill bit to centre it on the
pilot hole (photo E).

F

10 Draw round the template to mark the
hole in the head lining (photo F).

G

11 Using a sharp hobby knife, cut out the
marked hole in the head lining. This
ensures that the hole saw will not rip
the head lining (photo G).

H

12 Fit the hole saw to the electric drill
(photo H).

I

13 Drill the hole for the vent (photo I).

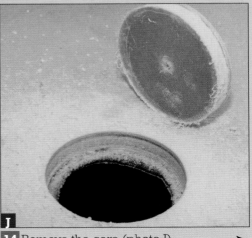

J

14 Remove the core (photo J).   ➤

15 View of the hole from inside (photo K).

16 Following the instructions supplied with the vent, fit the vent in place and mark the attachment holes (photo L).

17 Measure the length and the 'core' diameter of the attachment screws. The drill bit must be a little bigger than the core diameter, otherwise the fibreglass resin will shatter as you drive in the screws.

18 Wrap some masking tape around the drill bit to indicate the depth of the holes. You don't want these holes to go all the way through to the inside (photo M).

19 Drill the holes for the attachment screws.

20 Put a little sealant into each hole to prevent water penetrating the core. This is very important if the core is made of balsa wood because it can easily rot (photo N).

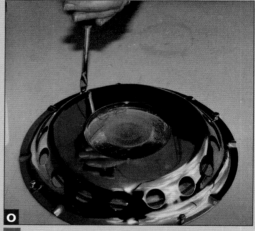

21 Drive in all the attachment screws (photo O).

22 The finished job (photo P).

63 Replacing a stanchion base

WHY IT NEEDS TO BE DONE

Sometimes it's necessary to replace a broken fitting or, as in this case, cure a leak which has started where the stanchion base is attached to the deck.

WHEN IT SHOULD BE DONE

As necessary, when a problem arises.

TOOLS NEEDED

Spanners and screwdrivers to suit, a hobby knife, sealant and masking tape.

1 Locate the nuts holding the fitting to the deck (photo A).

This can be difficult on some boats as quite a few modern builders use a moulded headliner, making access difficult. If you can't get behind the headliner, you'll have to cut an access panel (photo B).

To save cost, some modern builders also use an aluminium plate moulded into the deck into which the attachment bolts are screwed instead of using nuts. Stainless steel bolts screwed into aluminium plates

will cause corrosion of the plates and it may be very difficult to remove the bolts. The corrosion may also eat away the thread in the plates causing the bolts to become loose. It's likely that you'll have to use a nut when refitting the stanchion and access to the back of the plate may be very difficult. Here, an access panel has been provided (photo C).

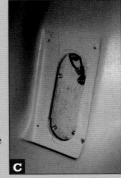

2 Here (photo D) the nuts were behind the headlining, which was glued in place. As the headlining had been attached behind wooden panelling, less damage would be caused by cutting the headlining rather than attempting to remove the panelling.

➤

E

3 Remove the nuts and washers. If you are working on your own, you can often stop the nut rotating by using a ring spanner, which is prevented from rotating when it comes up against the structure (photo E).

F

4 Remove the attachment bolts and the fitting (in this case the stanchion base) (photo F).

G

5 Clean up the deck, removing any old sealant. In this case it was found that Thinners Number 9 dissolved the sealant. You must make sure that whatever thinner you use does not damage the deck. Try a little on a hidden area first and do wear protective gloves or a barrier cream (photo G).

H

6 Remove and clean the backing plate (photo H).

I

J

7 Clean the underside of the deck with a scraper and abrasive paper (photos I & J).

K

8 Because the problem was a leak, the bolt holes were plugged using small softwood plugs (photo K) and the dehumidifier was left running to dry everything out (photo L).

L

9 Apply masking tape to the deck around where the fitting is to go. Ensure that the tape will not go under the fitting - leave a little gap (photo M).

10 When all is dry, apply the sealant to the deck putting a ring of sealant around the bolt holes. The aim is to keep the ends of the bolts free of sealant during assembly (photos N & O).

11 Apply sealant around the bolts just below the head, so that the end of the bolt will stay free of sealant when you push it in place (photo P).

12 Assemble the fitting and bolts, holding the fitting just clear of the deck, trying to prevent the sealant being squashed (photo Q).

13 Now bed down the fitting (photo R).
14 Put the backing plate in place, followed by the washers and the nuts (photo S).

15 Tighten the attachment bolts and nuts. This will almost certainly require two pairs of hands to ensure that the screwdriver slots in the bolt heads remain parallel to the gunwhale or fitting edge as befits a job well done. Some screwdrivers allow a spanner to be used on their shank to get a good grip (photo T).

16 When the sealant is cured, trim it to the fitting (photo U).
17 Remove the masking tape (photo V).
18 Replace any interior fittings that were removed to gain access.
19 Headlining can be glued back in place using 'Copydex', following the instructions on the container (photo W).

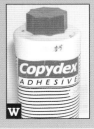

LEAKS

64 Locating a leak

WHY IT NEEDS TO BE DONE
Because you find dampness inside the boat.

WHEN IT SHOULD BE DONE
If you suspect a leak.

SKILL LEVEL
It depends how much you have to take apart. Perseverance is often more important than skill.

TOOLS NEEDED
Unknown when you start. You may need to remove headlining or get inside cupboards.

If you find dampness inside the boat, it doesn't have to be caused by a leak. Condensation is the first thing to suspect.

Condensation
It's all too easy to have insufficient ventilation inside a boat, especially in the spring and autumn.

If there's condensation on the windows and frames, dampness may be expected on any cool surface. As it builds up, water will drip off onto whatever is below, maybe starting a false trail. If there's visible condensation, then there may well be condensation under mattresses, and behind seat cushions. More ventilation is the real cure, but proprietary underlays can be obtained which may remove the symptoms.

A real leak
Once you have eliminated condensation as the cause of any dampness, then you have to assume a leak.

1. Leaks may be very difficult to locate. The leak may not be where the dampness becomes visible.
2. Locate the nearest deck fitting that is level with or above where you found dampness inside the boat.
3. Examine the fastenings and any bedding compound for signs of movement, on the off-chance that the leak might be obvious.
4. Gain access to the hull / under-deck located close to the fitting and examine for signs of dampness or staining.
5. If you find nothing, you'll have to work upwards and outwards away from the damp area until you find the culprit.
6. If nothing becomes obvious, you'll need to dry everything thoroughly and then puff some talcum powder or French chalk onto the under surface, in the hope that it will allow the leak to be traced next time water comes in.
7. Try spraying the deck area with a powerful hose in the hope of generating the leak.
8. Dried out bedding compound around a hatch or window surround is often the cause. If this is the case, the only sure way of tackling the problem is to remove the frame and clean all parts very thoroughly, removing all traces of old sealant and dirt. Re-bed the frame using proper bedding compound. Trying to plaster some new sealant around the leaking frame is almost sure to be doomed to failure.

Water in the bilge

The first thing to check is whether it's salt or fresh water.

Fresh water

- Fresh water is likely to be from the fresh water system, although it may of course be from a hull / deck / window leak.
- If you've a hot water system, there will be a pressure relief valve on the tank, which may operate each time you heat up the tank. This will allow water into the bilge. Some builders pipe this to the main bilge, others fit no pipe at all (photo A).

- If this is your problem, lead the pipe from the relief valve to a bottle, which you can empty as necessary (photos B & C).
- Fresh water containing antifreeze will obviously be from the engine's fresh water cooling system. Check this.

Salt water

- Salt water will be from the seawater cooling system, a hull leak or the anchor locker.
- On the cooling system, a salt water leak is likely to leave traces of salt crystals at the site of the leak. Check especially the seawater pump, all the hoses and the exhaust injection bend.
- All but dripless sterngland seals will drip a little. Excess leakage indicates the need for some attention.
- A loose propshaft 'P' bracket may allow water to leak into the hull. Clean up all round the internal site of the 'P' bracket and dust with talc to find the leak.
- The bearing / water seal at the top of an internal rudder stock may be worn, allowing water in (photo D). I've known one make of boat where the oversized engine caused the boat to squat so much at high power that water entered the boat

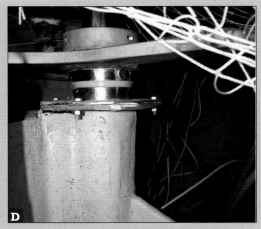

through the top of the rudder bearing.
- The 'O' ring seal in the paddlewheel log impeller may be leaking.
- If the anchor locker drain is blocked, seawater may find its way into the bilge. Empty the locker and hose some water in to see what happens.
- Check for weeping around the keel bolts.

WINCHES

65 Winch servicing

WHY IT NEEDS TO BE DONE
Grease can congeal in use, reducing the power of a winch, causing ratchets to stick and the winch to release without warning. This can cause injury.

WHEN IT SHOULD BE DONE
Cruising yachts need to have their winches serviced annually, racing yachts several times per season.

You don't need many tools or much skill, but a bit of patience may be helpful when reassembling the pawls (that stop the winch running backwards). Do resist the temptation to plaster everything with grease!

TOOLS NEEDED
The instructions that came with the winch will be helpful, as will a list of the tools required.
Petrol for cleaning, winch grease and light oil, lintless rag and a small soft paintbrush. A winch service kit will provide spares should they be required.
A container to put all the bits in.

Difficulty Rating: ▣▣□□□

Winches tend to be mounted in places where, if a bit drops off, it will roll into the water, so take your time and keep hold of everything. Put items down in the cockpit, not on the deck or coaming.

The illustrations show how to service an Andersen 40ST self-tailing winch (photo A), but others are similar.

B2　**B3**

A

1 Remove the three screws (photo B1) in the top of the winch self-tailing arm and then the top cap (photo B2). Mark the self-tailing arm and the top face of the base (photo B3) so

that the arm will go back in the correct position on reassembly.
2 Remove the self-tailing arm together with the drum. As the drum comes clear, the arm can fall off so hold the unit with thumb and finger either side of the arm (photo C).

B1

C

D

F1

F2

3 Insert the winch handle into the socket in the drive shaft. Pull the drive shaft gently up using the handle; at the same time turn both the upper and lower gear-wheels slowly in a clockwise direction (photo E1)...

E1

the drive shaft (photo F1) so that the lower pawls can be slid out of their housing (photo F2), bringing their springs with them (photo F3). The spiral springs of the upper pawls can also be removed using the small screwdriver (photo F4). Do not lose these springs and make sure they don't jump out as they are freed (photo F5). It's definitely advisable to carry out this operation away from the sides of the boat.

E2 E3 E4

F3

F4

...and remove it completely (photos E2, E3 & E4). This operation needs a finger of one hand on each of the gears to do the clockwise turning while the other exerts a gentle upward pressure without rotation on the handle. A co-ordinated 'softly softly' approach will work where force will get you nowhere.

4 Remove the lower pawls, arm springs and springs. The upper pawls aren't removable. Use the blade of a small screwdriver to move the ratchet springs gently from their grooves in

F5

G1

G2

I3

I4

5 Release the roller bearing by gently inserting a small screwdriver between the roller bearing and the base and then carefully removing it. Keep the cage of the roller bearing level or it will stick. A little bit of patience will pay off here (photos G1 & G2).

7 Use a screwdriver to remove the shaft (photos I1 & I2) and gearwheel (photo I3). The shaft will be slippery so don't drop it as it comes free. Use a screwdriver to lift the lower ratchet gear so that it can be removed (photo I4).

8 Clean the balls at the top of the post with a rag. DO NOT remove these balls OR the small retaining ring above them. Clean all parts in petrol, including the gear teeth, drive shaft, roller bearing track and ball track inside the drum (photos J1, J2, J3 & J4).

H1

H2

6 Remove the bush (photo H1) and ratchet gear (photo H2).

I1

I2

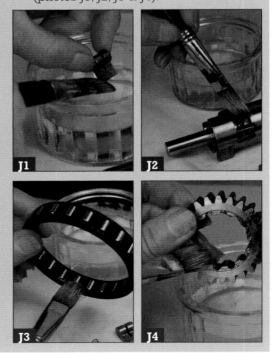
J1

J2

J3

J4

K1 **K2**

9 Assemble in reverse order.
(photos K1 & K2).

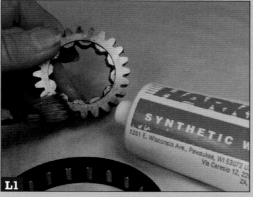

L1

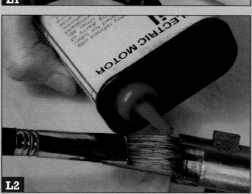

L2

10 During assembly, LIGHTLY grease the gear teeth, drive shaft, roller bearing, shaft, balls, pawls, springs and composite bushings, using a small soft brush. Pawls may be lubricated with either a VERY THIN film of WINCH grease or with light machine oil (photos L1 & L2).

Note:

It's very important that the pawls can move freely. Before inserting the drive shaft into the base, check the functioning of each pawl by pushing the pawl against its spring. The pawls should move smoothly and automatically return to their normal position, where the pawls engage with the teeth. If the pawls do not function correctly, clean and lubricate the pawls once more. If they still don't work correctly, replace the springs with new ones (photo M).

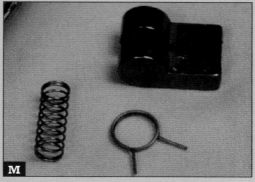

M

Incorrectly-functioning pawls may lead to unexpected release of the winch.

Photo N shows the workings of a Lewmar winch.

N

WINDLASSES

66 Routine maintenance

WHY IT NEEDS TO BE DONE

The windlass sits right at the front of the boat and is subject to regular dousing with salt water. If the electric motor is in the chain locker it is in a damp salt-laden environment. Although designed for the marine environment, corrosion must be likely.

WHEN IT SHOULD BE DONE

Operate the windlass regularly and hose down with fresh water. Give it a thorough annual check, as detailed in the handbook.

TOOLS NEEDED

Grease, grease gun and open-ended spanners.

Difficulty Rating: ▮ ☐ ☐ ☐ ☐

(Rating will be 3 if the windlass is old and has not had any annual maintenance - in fact taking it apart may not even be possible.)

Most of the illustrations are of an SL 555 Sea Tiger manual windlass, but the principles will be similar for many other units. The electric windlass illustrated later is a Lofrans electric capstan.

Windlasses may be electrically operated, or manual. Strictly speaking a windlass has a horizontal drum. If the drum is vertical, it should be called a capstan.

- It's essential that you hose down your windlass with fresh water frequently and also exercise it (photos A1, A2 & A3).

- Regularly treat any grease nipples to a visit from a grease gun. Also, oil the spindles and the main shaft (sparingly) where they enter the case (photo E).

- Always make off the anchor cable to a strong point on the foredeck. Any severe 'snubbing' on the windlass will cause damage and eventual failure to shafts, bearings or gears. Make sure that the free end of the chain is on top of the cleat so that it can be released (photo B).

- Try to use a nylon rope strop so that there's some 'give' in the system. This damps down loading. You can use a rolling hitch (photo C)...

retaining pin

- ...or a 'chain hook' to attach the strop (photo D).

- Remember to fit the retaining pin to stop the chain jumping out and sawing off the bows! (Photo F.)
- You may also want to haul the chain by hand, so a chain stopper is recommended (photos G1 & G2).

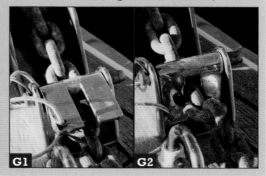

ANNUAL MAINTENANCE

Before you start to take things apart, check for any play in the main shaft bearings by exerting a sideways force on the gypsy and drum. Play indicates wear, which will allow water to penetrate into the casing, causing corrosion of the mechanism.

Re-bushing will require stripping of the windlass, best left to the manufacturer or their agent.

Stripper

1 Remove the bolt securing the 'rope/chain stripper'. As this is a stainless steel bolt in the aluminium casing there will possibly be corrosion. Regular loosening and greasing of this bolt is essential. Another way of keeping this corrosion at bay might be to use 'Duralac' on assembly. This is usually used on permanent joints to insulate aluminium from stainless steel, but I feel that this will make a joint that is easier to undo than a corroded one (photo H).

2 Now remove the 'stripper'. Corrosion is the enemy here as well and if you can't remove the stripper, you won't be able to remove the gypsy. However, you can do most of the essential maintenance with the gypsy in place, so if carried out annually, you'll be able to keep the windlass running for quite a few years. If you really need to remove the stripper, you'll have to remove the windlass from the deck so that, if necessary,

you can drill out the bolt out and drive the stripper through from the bottom (photo I).

Pawl

3 Engage the locking pawl (photo J).
4 Remove the warping drum retaining bolt (photo K).

Warping drum

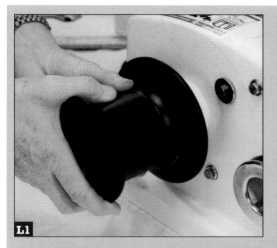

L1

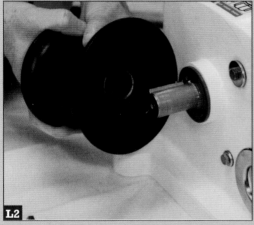

L2

L3

5 Slide the drum off the main shaft. Unless this is done at least annually, the chances of the drum seizing onto the shaft are high, especially if the drum is aluminium. (Photos L1, L2 & L3.)

Key

M

6 Make sure you don't lose the key (photo M).

N1

N2

Clutch nut

N3

7 Remove the clutch nut (photos N1, N2 & N3). ➤

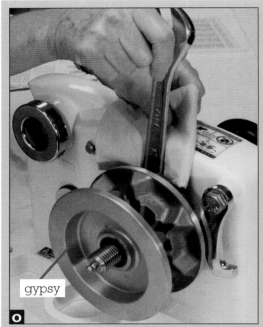

gypsy

O

8 Protect the body of the windlass and lever the gypsy off the clutch cone (photo O)…

P1

P2

9 …and remove the gypsy (photos P1 & P2).

- **Note:** Grease enters the grease nipple (photo Q1) and exits through the holes in the thread and bearing of the main shaft, to lubricate the clutch nut, gypsy bearing and clutch (photo Q2).

grease nipple

Q1

exit holes

Q2

10 If it's necessary to remove the clutch cone, you'll have to drive out the roll pin though this hole (photo R).

clutch

R

S1

S2

11 If you suspect that there's water in the casing, you'll need to remove the windlass from the deck, so that you can remove the base plate. (Photos S1,S2 & S3).

S3

- Don't damage the gasket. Clean the gears and re-grease, but if there's much corrosion have the windlass refurbished, because the ratchet springs may break and the ratchets may seize.

- Remember to fit the nylon insulating bushes when you refit the windlass, to keep the stainless steel bolts from contacting the aluminium casing (photo T). If they've split, do fit new ones. After you've used the windlass a few times, recheck the tightness of the securing bolts.

brushes

T

ELECTRIC WINDLASSES

If you have an electric windlass, you must keep your eyes open for corrosion of the motor casing and the electric terminals.

- This motor is mounted in the chain locker and the condensation is very apparent (photo U).

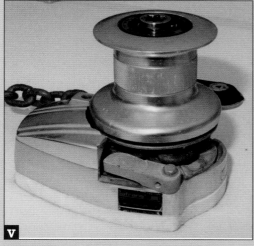

- A vertical electric windlass (capstan) (photo V).

- Its pawl pivots on a stainless steel bolt screwed into the aluminium base. Although there is a nylon spacer where the bolt passes through the pawl, the stainless thread will eventually seize in the aluminium casting. Remove this bolt annually and coat the thread with 'Duralac' paste to delay corrosion. Unless you do this you will not be able to remove the gypsy if you need to.

- Keep all the electrical connections tight and free of corrosion. The current is very high and corrosion will cause severe voltage drops (photo W).

INFLATABLES

67 Puncture repair

WHY IT NEEDS TO BE DONE
If the inflatable has a leak.

WHEN IT SHOULD BE DONE
When a leak is discovered.

TOOLS NEEDED
Sandpaper, cloth, degreasing agent
(photo A), paintbrush, repair kit,
spoon, scissors.

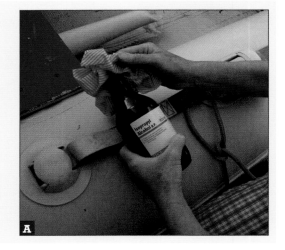

A

Normally, inflatables are made from
'HYPERLON' or 'PVC'.

It's essential that the correct adhesive and
patch is used according to the material.
Using the manufacturer's own repair kit
will ensure the correct materials (photo
B). Other kits are satisfactory as long as
they match the material.

The following is the procedure for a
Zodiac inflatable, but the principles are
the same for all. Make sure that you follow
the manufacturer's instructions or the
repair is likely to fail. Especially follow the
recommendations on temperature and
humidity.

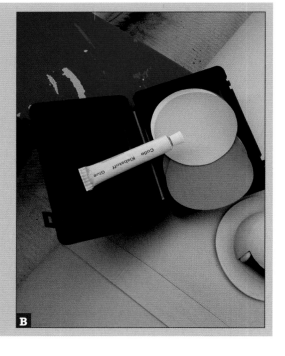

B

Difficulty Rating:

First find the leak

1 The leak may be obvious by sight or sound.

2 If not, use a brush and a solution of washing up liquid in water (photo C).

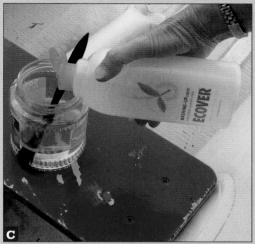

C

E

3 Brush the solution onto the inflation valves, and the seams (photo D).

4 Brush the solution onto likely areas of damage and, if necessary, the whole dinghy (photo E).

5 A leak will cause bubbles to form as the air escapes (photo F).

6 Replacement valves should be available from your supplier.

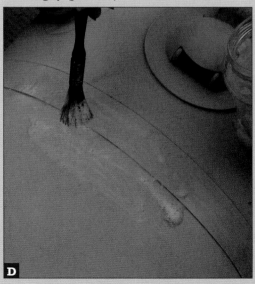

D

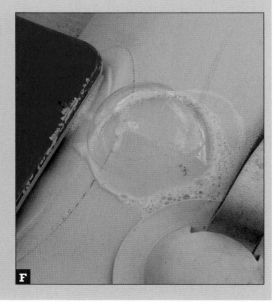

F

Repairing the leak

1 Deflate the dinghy fully.

2 Put a flat firm surface, such as a piece of plywood, under the damaged area.

3 Clean the damaged area with an alcohol-based degreaser (medical alcohol or methylated spirit) (photo G).

4 Choose a patch that covers an area at least 50mm from the leak, and mark the area with a pencil (photo H).

5 Roughen the surface of the patch and the dinghy with abrasive paper (photo I).

6 Remove any dust from the damage site and the patch with alcohol and allow to dry for 5 minutes. Apply 3 thin coats of adhesive to BOTH patch and dinghy, allowing 5 minutes between each coat (photos J & K).

8 Allow a further 5 minutes drying time, then place the patch lightly in place and position accurately (photo L).

9 Working from the centre outwards,

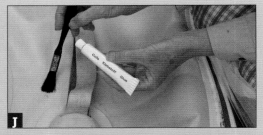

remove air bubbles and press the patch firmly into place using a spoon (photo M).

10 Clean off any surplus adhesive with a solvent.

11 Do not expose the repair to sun or rain during the repair or until the adhesive has cured.

12 Allow at least 24 hours before inflating the dinghy (photo N).

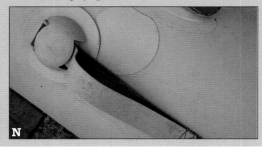